Delmar's Test Preparation Series

D1429681

Medium/Heavy Truck Test

Electrical/Electronic Systems (Test T6)
Second Edition

THOMSON

DELMAR LEARNING

Australia Canada Mexico Singapore Spain United Kingdom United States

THOMSON

DELMAR LEARNING

Delmar's ASE Test Preparation Series
ASE Medium/Heavy Truck Test T6 (Electrical/Electronic Systems)

Business Unit Director:
Alar Elken

Executive Editor:
Sandy Clark

Acquisitions Editor:
San Rao

Editorial Assistant:
Bryan Viggiani

**Business and Product
Development Consultant:**
David Koontz

Developmental Editor:
Christopher Shortt

Executive Marketing Manager:
Maura Theriault

Marketing Coordinator:
Brian McGrath

Channel Manager:
Fair Huntoon

Executive Production Manager:
Mary Ellen Black

Production Manager:
Larry Main

Production Editor:
Tom Stover

Cover Designer:
Michael Egan

ISBN: 1-7668-4896-5

NOTICE TO THE READER

Publisher does not warrant or guarantee any of the products described herein or perform any independent analysis in connection with any of the product information contained herein. Publisher does not assume, and expressly disclaims, any obligation to obtain and include information other than that provided to it by the manufacturer.

The reader is expressly warned to consider and adopt all safety precautions that might be indicated by the activities herein and to avoid all potential hazards. By following the instructions contained herein, the reader willingly assumes all risks in connection with such instructions.

The Publisher makes no representation or warranties of any kind, including but not limited to, the warranties of fitness for particular purpose or merchantability, nor are any such representations implied with respect to the material set forth herein, and the publisher takes no responsibility with respect to such material. The publisher shall not be liable for any special, consequential, or exemplary damage resulting, in whole or part, from the readers' use of, or reliance upon, this material.

Contents

Section 3 Are You Sure You're Ready for Test T6?

Section 4 An Overview of the System

Section 5 Sample Test for Practice

Section 6 Additional Test Questions for Practice

Section 7 Appendices

Preface

This book is just one of a comprehensive series designed to prepare technicians to take and pass every ASE test. Delmar's series covers all of the Automotive tests A1 through A8 as well as Advanced Engine Performance L1 and Parts Specialist P2. The series also covers the five Collision Repair tests and the eight Medium/Heavy Duty truck tests.

Before any book in this series was written, Delmar staff met with and surveyed technicians and shop owners who have taken ASE tests and have used other preparatory materials. We found that they wanted, first and foremost, *lots* of practice tests and questions. Each book in our series contains a general knowledge pretest, a sample test, and additional practice questions. You will be hard-pressed to find a test prep book with more questions for you to practice with. We have worked hard to ensure that these questions match the ASE style in types of questions, quantities, and level of difficulty.

Technicians also told us that they wanted to understand the ASE test and to have practical information about what they should expect. We have provided that as well, including a history of ASE and a section devoted to helping the technician "Take and Pass Every ASE Test" with case studies, test-taking strategies, and test formats.

Finally, techs wanted refresher information and reference. Each of our books includes an overview section that is referenced to the task list. The complete task lists for each test appear in each book for the user's reference. There is also a complete glossary of terms for each booklet.

So whether you're looking for a sample test and a few extra questions to practice with or a complete introduction to ASE testing, with support for preparing thoroughly, this book series is an excellent answer.

We hope you benefit from this book and that you pass every ASE test you take!

Your comments, both positive and negative, are certainly encouraged! Please contact us at:

Automotive Editor
Delmar Learning
Executive Woods
5 Maxwell Drive
Clifton Park, NY 12065-2919

1 The History of ASE

History

Originally known as The National Institute for Automotive Service Excellence (NIASE), today's ASE was founded in 1972 as a nonprofit, independent entity dedicated to improving the quality of automotive service and repair through the voluntary testing and certification of automotive technicians. Until that time, consumers had no way of distinguishing between competent and incompetent automotive technicians. In the mid-1960s and early 1970s, efforts were made by several automotive industry affiliated associations to respond to this need. Though the associations were nonprofit, many regarded certification test fees merely as a means of raising additional operating capital. Also, some associations, having a vested interest, produced test scores heavily weighted in the favor of their members.

NIASE

From these efforts a new independent, nonprofit association, the National Institute for Automotive Service Excellence (NIASE), was established much to the credit of two educators, George R. Kinsler, Director of Program Development for the Wisconsin Board of Vocational and Adult Education in Madison, WI, and Myron H. Appel, Division Chairman at Cypress College in Cypress, CA.

Early efforts were to encourage voluntary certification in four general areas:

TEST AREA	TITLES
I. Engine	Engines, Engine Tune-Up, Block Assembly, Cooling and Lube Systems, Induction, Ignition, and Exhaust
II. Transmission	Manual Transmissions, Driveline and Rear Axles, and Automatic Transmissions
III. Brakes and Suspension	Brakes, Steering, Suspension, and Wheels
IV. Electrical/Air Conditioning	Body/Chassis, Electrical Systems, Heating, and Air Conditioning

In early NIASE tests, Mechanic A, Mechanic B type questions were used. Over the years the trend has not changed, but in mid-1984 the term was changed to Technician A, Technician B to better emphasize sophistication of the skills needed to perform successfully in the modern motor vehicle industry. In certain tests the term used is Estimator A/B, Painter A/B, or Parts Specialist A/B. At about that same time, the logo was changed from "The Gear" to "The Blue Seal," and the organization adopted the acronym ASE for Automotive Service Excellence.

Since those early beginnings, several other related trades have been added. ASE now administers a comprehensive series of certification exams for automotive and light truck repair technicians, medium and heavy truck repair technicians, alternate fuels technicians, engine machinists, collision repair technicians, school bus repair technicians, and parts specialists.

The Series and Individual Tests

- Automotive and Light Truck Technician; consisting of: Engine Repair—Automatic Transmission/Transaxle—Manual Drive Train and Axles—Suspension and Steering—Brakes—Electrical/Electronic Systems—Heating and Air Conditioning—Engine Performance
- Medium and Heavy Truck Technician; consisting of: Gasoline Engines—Diesel Engines—Drive Train—Brakes—Suspension and Steering—Electrical/Electronic Systems—HVAC—Preventive Maintenance Inspection
- Alternate Fuels Technician; consisting of: Compressed Natural Gas Light Vehicles
- Advanced Series; consisting of: Automobile Advanced Engine Performance and Advanced Diesel Engine Electronic Diesel Engine Specialty
- Collision Repair Technician; consisting of: Painting and Refinishing—Non-Structural Analysis and Damage Repair—Structural Analysis and Damage Repair—Mechanical and Electrical Components—Damage Analysis and Estimating
- Engine Machinist Technician; consisting of: Cylinder Head Specialist—Cylinder Block Specialist—Assembly Specialist
- School Bus Repair Technician; consisting of: Body Systems and Special Equipment—Drive Train—Brakes—Suspension and Steering—Electrical/Electronic Systems—Heating and Air Conditioning
- Parts Specialist; consisting of: Automobile Parts Specialist—Medium/Heavy Truck Parts Specialist

A Brief Chronology

1970–1971	Original questions were prepared by a group of forty auto mechanics teachers from public secondary schools, technical institutes, community colleges, and private vocational schools. These questions were then professionally edited by testing specialists at Educational Testing Service (ETS) at Princeton, New Jersey, and thoroughly reviewed by training specialists associated with domestic and import automotive companies.
1971	July: About eight hundred mechanics tried out the original test questions at experimental test administrations.
1972	November and December: Initial NIASE tests administered at 163 test centers. The original automotive test series consisted of four tests containing eighty questions each. Three hours were allotted for each test. Those who passed all four tests were designated Certified General Auto Mechanic (GAM).
1973	April and May: Test 4 was increased to 120 questions. Time was extended to four hours for this test. There were now 182 test centers. Shoulder patch insignias were made available.

	November: Automotive series expanded to five tests. Heavy-Duty Truck series of six tests introduced.
1974	November: Automatic Transmission (Light Repair) test modified. Name changed to Automatic Transmission.
1975	May: Collision Repair series of two tests is introduced.
1978	May: Automotive recertification testing is introduced.
1979	May: Heavy-Duty Truck recertification testing is introduced.
1980	May: Collision Repair recertification testing is introduced.
1982	May: Test administration providers switched from Educational Testing Service (ETS) to American College Testing (ACT). Name of Automobile Engine Tune-Up test changed to Engine Performance test.
1984	May: New logo was introduced. ASE's "The Blue Seal" replaced NIASE's "The Gear." All reference to Mechanic A, Mechanic B was changed to Technician A, Technician B.
1990	November: The first of the Engine Machinist test series was introduced.
1991	May: The second test of the Engine Machinist test series was introduced. November: The third and final Engine Machinist test was introduced.
1992	May: Name of Heavy-Duty Truck Test series changed to Medium/Heavy Truck test series.
1993	May: Automotive Parts Specialist test introduced. Collision Repair expanded to six tests. Light Vehicle Compressed Natural Gas test introduced. Limited testing begins in English-speaking provinces of Canada.
1994	May: Advanced Engine Performance Specialist test introduced.
1996	May: First three tests for School Bus Technician test series introduced. November: A Collision Repair test is added.
1997	May: A Medium/Heavy Truck test is added.
1998	May: A diesel advanced engine test is introduced: Electronic Diesel Engine Diagnosis Specialist. A test is added to the School Bus test series.

By the Numbers

Following are the approximate number of ASE technicians currently certified by category. The numbers may vary from time to time but are reasonably accurate for any given period. More accurate data may be obtained from ASE, which provides updates twice each year, in May and November after the Spring and Fall test series.

There are more than 338,000 Automotive Technicians with over 87,000 at Master Technician (MA) status. There are 47,000 Truck Technicians with over 19,000 at Master Technician (MT) status. There are 46,000 Collision Repair/Refinish Technicians with 7,300 at Master Technician (MB) status. There are 1,200 Estimators. There are 6,700 Engine Machinists with over 2,800 at Master Machinist Technician (MM) status. There are also 28,500 Automobile Advanced Engine Performance Technicians and over 2,700 School Bus Technicians for a combined total of more than 403,000 Repair Technicians. To this number, add over 22,000 Automobile Parts Specialists, and over 2,000 Truck Parts Specialists for a combined total of over 24,000 parts specialists.

There are over 6,400 ASE Technicians holding both Master Automotive Technician and Master Truck Technician status, of which 350 also hold Master Body Repair status. Almost 200 of these Master Technicians also hold Master Machinist status and five Technicians are certified in all ASE specialty areas.

Almost half of ASE certified technicians work in new vehicle dealerships (45.3 percent). The next greatest number work in independent garages with 19.8 percent. Next is tire dealerships with 9 percent, service stations at 6.3 percent, fleet shops at 5.7 percent, franchised volume retailers at 5.4 percent, paint and body shops at 4.3 percent, and specialty shops at 3.9 percent.

Of over 400,000 automotive technicians on ASE's certification rosters, almost 2,000 are female. The number of female technicians is increasing at a rate of about 20 percent each year. Women's increasing interest in the automotive industry is further evidenced by the fact that, according to the National Automobile Dealers Association (NADA), they influence 80 percent of the decisions of the purchase of a new automobile and represent 50 percent of all new car purchasers. Also, it is interesting to note that 65 percent of all repair and maintenance service customers are female.

The typical ASE certified technician is 36.5 years of age, is computer literate, deciphers a half-million pages of technical manuals, spends one hundred hours per year in training, holds four ASE certificates, and spends about $27,000 for tools and equipment. Twenty-seven percent of today's skilled ASE certified technicians attended college, many having earned an Associate of Science degree in Automotive Technology.

ASE

ASE's mission is to improve the quality of vehicle repair and service in the United States through the testing and certification of automotive repair technicians. Prospective candidates register for and take one or more of ASE's thirty-three exams. The tests are grouped into specialties for automobile, medium/heavy truck, school bus, and collision repair technicians as well as engine machinists, alternate fuels technicians, and parts specialists.

Upon passing at least one exam and providing proof of two years of related work experience, the technician becomes ASE certified. A technician who passes a series of exams earns ASE Master Technician status. An automobile technician, for example, must pass eight exams for this recognition.

The tests, conducted twice a year at over seven hundred locations around the country, are administered by American College Testing (ACT). They stress real-world diagnostic and repair problems. Though a good knowledge of theory is helpful to the technician in answering many of the questions, there are no questions specifically on theory. Certification is valid for five years. To retain certification, the technician must be retested to renew his or her certificate.

The automotive consumer benefits because ASE certification is a valuable yardstick by which to measure the knowledge and skills of individual technicians, as well as their commitment to their chosen profession. It is also a tribute to the repair facility employing ASE certified technicians. ASE certified technicians are permitted to wear blue and white ASE shoulder insignia, referred to as the "Blue Seal of Excellence," and carry credentials listing their areas of expertise. Often employers display their technicians' credentials in the customer waiting area. Customers look for facilities that display ASE's Blue Seal of Excellence logo on outdoor signs, in the customer waiting area, in the telephone book (Yellow Pages), and in newspaper advertisements.

The tests stress repair knowledge and skill. All test takers are issued a score report. In order to earn ASE certification, a technician must pass one or more of the exams and present proof of two years of relevant hands-on work experience. ASE certifications are valid for five years, after which time technicians must retest in order to keep up with changing technology and to remain in the ASE program. A nominal registration and test fee is charged.

To become part of the team that wears ASE's Blue Seal of Excellence®, please contact:

National Institute for Automotive Service Excellence
13505 Dulles Technology Drive
Herndon, VA 20171-3421

2 Take and Pass Every ASE Test

ASE Testing

Participating in an Automotive Service Excellence (ASE) voluntary certification program gives you a chance to show your customers that you have the "know-how" needed to work on today's modern vehicles. The ASE certification tests allow you to compare your skills and knowledge to the automotive service industry's standards for each specialty area.

If you are the "average" automotive technician taking this test, you are in your mid-thirties and have not attended school for about fifteen years. That means you probably have not taken a test in many years. Some of you, on the other hand, have attended college or taken postsecondary education courses and may be more familiar with taking tests and with test-taking strategies. There is, however, a difference in the ASE test you are preparing to take and the educational tests you may be accustomed to.

Who Writes the Questions?

The questions on an educational test are generally written, administered, and graded by an educator who may have little or no practical hands-on experience in the test area. The questions on all ASE tests are written by service industry experts familiar with all aspects of the subject area. ASE questions are entirely job-related and designed to test the skills that you need to know on the job.

The questions originate in an ASE "item-writing" workshop where service representatives from domestic and import automobile manufacturers, components and equipment manufacturers, and vocational educators meet in a workshop setting to share their ideas and translate them into test questions. Each test question written by these experts is reviewed by all of the members of the group. The questions deal with the practical problems of diagnosis and repair that are experienced by technicians in their day-to-day hands-on work experiences.

All of the questions are pretested and quality-checked in a nonscoring section of tests by a national sample of certifying technicians. The questions that meet ASE's high standards of accuracy and quality are then included in the scoring sections of future tests. Those questions that do not pass ASE's stringent tests are sent back to the workshop or are discarded. ASE's tests are monitored by an independent proctor and are administered and machine-scored by an independent provider, American College Testing (ACT). All ASE tests have a three-year revision cycle.

Testing

If you think about it, we are actually tested on about everything we do. As infants, we were tested to see when we could turn over and crawl, later when we could walk or talk. As adolescents, we were tested to determine how well we learned the material presented in school and in how we demonstrated our accomplishments on the athletic field. As working adults, we are tested by our supervisors on how well we have completed an assignment or project. As nonworking adults, we are tested by our family on everyday activities, such as housekeeping or preparing a meal. Testing, then, is one of those facts of life that begins in the cradle and follows us to the grave.

Testing is an important fact of life that helps us to determine how well we have learned our trade. Also, tests often help us to determine what opportunities will be available to us in the future. To become ASE certified, we are required to take a test in every subject in which we wish to be recognized.

Be Test-Wise

In spite of the widespread use of tests, most technicians are not very test-wise. An ability to take tests and score well is a skill that must be acquired. Without this knowledge, the most intelligent and prepared technician may not do well on a test.

We will discuss some of the basic procedures necessary to follow in order to become a test-wise technician. Assume, if you will, that you have done the necessary study and preparation to score well on the ASE test.

Different approaches should be used for taking different types of tests. The different basic types of tests include: essay, objective, multiple-choice, fill-in-the-blank, true-false, problem solving, and open book. All ASE tests are of the four-part multiple-choice type.

Before discussing the multiple-choice type test questions, however, there are a few basic principles that should be followed before taking any test.

Before the Test

Do not arrive late. Always arrive well before your test is scheduled to begin. Allow ample time for the unexpected, such as traffic problems, so you will arrive on time and avoid the unnecessary anxiety of being late.

Always be certain to have plenty of supplies with you. For an ASE test, three or four sharpened soft lead (#2) pencils, a pocket pencil sharpener, erasers, and a watch are all that are required.

Do not listen to pretest chatter. When you arrive early, you may hear other technicians testing each other on various topics or making their best guess as to the probable test questions. At this time, it is too late to add to your knowledge. Also the rhetoric may only confuse you. If you find it bothersome, take a walk outside the test room to relax and loosen up.

Read and listen to all instructions. It is important to read and listen to the instructions. Make certain that you know what is expected of you. Listen carefully to verbal instructions and pay particular attention to any written instructions on the test paper. Do not dive into answering questions only to find out that you have answered the wrong question by not following instructions carefully. It is difficult to make a high score on a test if you answer the wrong questions.

These basic principles have been violated in almost every test ever given. Try to remember them. They are essential for success.

Objective Tests

A test is called an objective test if the same standards and conditions apply to everyone taking the test and there is only one correct answer to each question. Objective tests primarily measure your ability to recall information. A well-designed objective test can also test your ability to understand, analyze, interpret, and apply your knowledge. Objective tests include true-false, multiple-choice, fill-in-the-blank, and matching questions.

Objective questions, not generally encountered in a classroom setting, are frequently used in standardized examinations. Objective tests are easy to grade and also reduce the amount of paperwork necessary to administer. The objective tests are used in entry-level programs or when very large numbers are being tested. ASE's tests consist exclusively of four-part multiple-choice objective questions in all of their tests.

Taking an Objective Test

The principles of taking an objective test are somewhat different from those used in other types of tests. You should first quickly look over the test to determine the number of questions, but do not try to read through all of the questions. In an ASE test, there are usually between forty and eighty questions, depending on the subject matter. Read through each question before marking your answer. Answer the questions in the order they appear on the test. Leave the questions blank that you are not sure of and move on to the next question. You can return to those unanswered questions after you have finished the others. They may be easier to answer at a later time after your mind has had additional time to consider them on a subconscious level. In addition, you might find information in other questions that will help you to answer some of them.

Do not be obsessed by the apparent pattern of responses. For example, do not be influenced by a pattern like **d**, **c**, **b**, **a**, **d**, **c**, **b**, **a** on an ASE test.

There is also a lot of folk wisdom about taking objective tests. For example, there are those who would advise you to avoid response options that use certain words such as *all*, *none*, *always*, *never*, *must*, and *only*, to name a few. This, they claim, is because nothing in life is exclusive. They would advise you to choose response options that use words that allow for some exception, such as *sometimes*, *frequently*, *rarely*, *often*, *usually*, *seldom*, and *normally*. They would also advise you to avoid the first and last option (A and D) because test writers, they feel, are more comfortable if they put the correct answer in the middle (B and C) of the choices. Another recommendation often offered is to select the option that is either shorter or longer than the other three choices because it is more likely to be correct. Some would advise you to never change an answer since your first intuition is usually correct.

Although there may be a grain of truth in this folk wisdom, ASE test writers try to avoid them and so should you. There are just as many **A** answers as there are **B** answers, just as many **D** answers as **C** answers. As a matter of fact, ASE tries to balance the answers at about 25 percent per choice **A**, **B**, **C**, and **D**. There is no intention to use "tricky" words, such as outlined above. Put no credence in the opposing words "sometimes" and "never," for example. When used in an ASE type question, one or both may be correct; one or both may be incorrect.

There are some special principles to observe on multiple-choice tests. These tests are sometimes challenging because there are often several choices that may seem possible, and it may be difficult to decide on the correct choice. The best strategy, in this case, is to first determine the correct answer before looking at the options. If you see the answer you decided on, you should still examine the options to make sure that none seem more correct than yours. If you do not know or are not sure of the answer, read each option very carefully and try to eliminate those options that you know to be wrong. That way, you can often arrive at the correct choice through a process of elimination.

If you have gone through all of the test and you still do not know the answer to some of the questions, then guess. Yes, guess. You then have at least a 25 percent chance of being correct. If you leave the question blank, you have no chance. In ASE tests, there is no penalty for being wrong. As the late President Franklin D. Roosevelt once advised a group of students, "It is common sense to take a method and try it. If it fails, admit it frankly and try another. But above all, try something."

During the Test

Mark your bubble sheet clearly and accurately. One of the biggest problems an adult faces in test-taking, it seems, is in placing an answer in the correct spot on a bubble sheet. Make certain that you mark your answer for, say, question 21, in the space on the bubble sheet designated for the answer for question 21. A correct response in the wrong bubble will probably be wrong. Remember, the answer sheet is machine scored and can only "read" what you have bubbled in. Also, do not bubble in two answers for the same question. For example, if you feel the answer to a particular question is **A** but think it may be **C,** do not bubble in both choices. Even if either **A** or **C** is correct, a double answer will score as an incorrect answer. It's better to take a chance with your best guess.

Review Your Answers

If you finish answering all of the questions on a test ahead of time, go back and review the answers of those questions that you were not sure of. You can often catch careless errors by using the remaining time to review your answers.

Don't Be Distracted

At practically every test, some technicians will invariably finish ahead of time and turn their papers in long before the final call. Do not let them distract or intimidate you. Either they knew too little and could not finish the test, or they were very self-confident and thought they knew it all. Perhaps they were trying to impress the proctor or other technicians about how much they know. Often you may hear them later talking about the information they knew all the while but forgot to respond on their answer sheet.

Use Your Time Wisely

It is not wise to use less than the total amount of time that you are allotted for a test. If there are any doubts, take the time for review. Any product can usually be made better with some additional effort. A test is no exception. It is not necessary to turn in your test paper until you are told to do so.

Don't Cheat

Some technicians may try to use a "crib sheet" during a test. Others may attempt to read answers from another technician's paper. If you do that, you are unquestionably assuming that someone else has a correct answer. You probably know as much, maybe more, than anyone else in the test room. Trust yourself. If you're still not convinced, think of the consequences of being caught. Cheating is foolish. If you are caught, you have failed the test.

Be Confident

The first and foremost principle in taking a test is that you need to know what you are doing, to be test-wise. It will now be presumed that you are a test-wise technician and are now ready for some of the more obscure aspects of test-taking.

An ASE-style test requires that you use the information and knowledge at your command to solve a problem. This generally requires a combination of information similar to the way you approach problems in the real world. Most problems, it seems, typically do not fall into neat textbook cases. New problems are often difficult to handle, whether they are encountered inside or outside the test room.

An ASE test also requires that you apply methods taught in class as well as those learned on the job to solve problems. These methods are akin to a well-equipped tool box in the hands of a skilled technician. You have to know what tools to use in a particular situation, and you must also know how to use them. In an ASE test, you will need to be able to demonstrate that you are familiar with and know how to use the tools.

You should begin a test with a completely open mind. At times, however, you may have to move out of your normal way of thinking and be creative to arrive at a correct answer. If you have diligently studied for at least one week prior to the test, you have bombarded your mind with a wide assortment of information. Your mind will be working with this information on a subconscious level, exploring the interrelationships among various facts, principles, and ideas. This prior preparation should put you in a creative mood for the test.

In order to reach your full potential, you should begin a test with the proper mental attitude and a high degree of self-confidence. You should think of a test as an opportunity to document how much you know about the various tasks in your chosen profession. If you have been diligently studying the subject matter, you will be able to take your test in serenity because your mind will be well organized. If you are confident, you are more likely to do well because you have the proper mental attitude. If, on the other hand, your confidence is low, you are bound to do poorly. It is a self-fulfilling prophecy.

Perhaps you have heard athletic coaches talk about the importance of confidence when competing in sports. Mental confidence helps an athlete to perform at the highest level and gain an advantage over competitors. Taking a test is much like an athletic event. You are competing against yourself, in a certain sense, because you will be trying to approach perfection in determining your answers. As in any competition, you should aim your sights high and be confident that you can reach the apex.

Anxiety and Fear

Many technicians experience anxiety and fear at the very thought of taking a test. Many worry, become nervous, and even become ill at test time because of the fear of failure. Many often worry about the criticism and ridicule that may come from their employer, relatives, and peers. Some worry about taking a test because they feel that the stakes are very high. Those who spent a great amount of time studying may feel they must get a high grade to justify their efforts. The thought of not doing well can result in unnecessary worry. They become so worried, in fact, that their reasoning and thinking ability is impaired, actually bringing about the problem they wanted to avoid.

The fear of failure should not be confused with the desire for success. It is natural to become "psyched-up" for a test in contemplation of what is to come. A little emotion can provide a healthy flow of adrenaline to peak your senses and hone your mental ability. This improves your performance on the test and is a very different reaction from fear.

Most technician's fears and insecurities experienced before a test are due to a lack of self-confidence. Those who have not scored well on previous tests or have no confidence in their preparation are those most likely to fail. Be confident that you will do well on your test and your fears should vanish. You will know that you have done everything possible to realize your potential.

Getting Rid of Fear

If you have previously experienced fear of taking a test, it may be difficult to change your attitude immediately. It may be easier to cope with fear if you have a better understanding of what the test is about. A test is merely an assessment of how much the technician knows about a particular task area. Tests, then, are much less threatening when thought of in this manner. This does not mean, however, that you should lower your self-esteem simply because you performed poorly on a test.

You can consider the test essentially as a learning device, providing you with valuable information to evaluate your performance and knowledge. Recognize that no one is perfect. All humans make mistakes. The idea, then, is to make mistakes before the test, learn from them, and avoid repeating them on the test. Fortunately, this is not as difficult as it seems. Practical questions in this study guide include the correct answers to consider if you have made mistakes on the practice test. You should learn where you went wrong so you will not repeat them in the ASE test. If you learn from your mistakes, the stage is set for future growth.

If you understood everything presented up until now, you have the knowledge to become a test-wise technician, but more is required. To be a test-wise technician, you not only have to practice these principles, you have to diligently study in your task area.

Effective Study

The fundamental and vital requirement to induce effective study is a genuine and intense desire to achieve. This is more basic than any rule or technique that will be given here. The key requirement, then, is a driving motivation to learn and to achieve.

If you wish to study effectively, first develop a desire to master your studies and sincerely believe that you will master them. Everything else is secondary to such a desire.

First, build up definite ambitions and ideals toward which your studies can lead. Picture the satisfaction of success. The attitude of the technician may be transformed from merely getting by to an earnest and energetic effort. The best direct stimulus to change may involve nothing more than the deliberate planning of your time. Plan time to study.

Another drive that creates positive study is an interest in the subject studied. As an automotive technician, you can develop an interest in studying particular subjects if you follow these four rules:

1. Acquire information from a variety of sources. The greater your interest in a subject, the easier it is to learn about it. Visit your local library and seek books on the subject you are studying. When you find something new or of interest, make inexpensive photocopies for future study.

2. Merge new information with your previous knowledge. Discover the relationship of new facts to old known facts. Modern developments in automotive technology take on new interest when they are seen in relation to present knowledge.

3. Make new information personal. Relate the new information to matters that are of concern to you. The information you are now reading, for example, has interest to you as you think about how it can help.

4. Use your new knowledge. Raise questions about the points made by the book. Try to anticipate what the next steps and conclusions will be. Discuss this new knowledge, particularly the difficult and questionable points, with your peers.

You will find that when you study with eager interest, you will discover it is no longer work. It is pleasure, and you will be fascinated by what you study. Studying can be like reading a novel or seeing a movie that overcomes distractions and requires no effort or willpower. You will discover that the positive relationship between interest and effort works both ways. Even though you perhaps began your studies with little or no interest, simply staying with it helped you to develop an interest in your studies.

Obviously, certain subject matter studies are bound to be of little or no interest, particularly in the beginning. Parts of certain studies may continue to be uninteresting. An honest effort to master those subjects, however, nearly always brings about some level of interest. If you appreciate the necessity and reward of effective studying, you will rarely be disappointed. Here are a few important hints for gaining the determination that is essential to carrying good conclusions into actual practice.

Make Study Definite

Decide what is to be studied and when it is to be studied. If the unit is discouragingly long, break it into two or more parts. Determine exactly what is involved in the first part and learn that. Only then should you proceed to the next part. Stick to a schedule.

The Urge to Learn

Make clear to yourself the relation of your present knowledge to your study materials. Determine the relevance with regard to your long-range goals and ambitions.

Turn your attention away from real or imagined difficulties as well as other things that you would rather be doing. Some major distractions are thoughts of other duties and of disturbing problems. These distractions can usually be put aside, simply shunted off by listing them in a notebook. Most technicians have found that by writing interfering thoughts down, their minds are freed from annoying tensions.

Adopt the most reasonable solution you can find or seek objective help from someone else for personal problems. Personal problems and worry are often causes of ineffective study. Sometimes there are no satisfactory solutions. Some manage to avoid the problems or to meet them without great worry. For those who may wish to find better ways of meeting their personal problems, the following suggestions are offered:

1. Determine as objectively and as definitely as possible where the problem lies. What changes are needed to remove the problem, and which changes, if any, can be made? Sometimes it is wiser to alter your goals than external conditions. If there is no perfect solution, explore the others. Some solutions may be better than others.

2. Seek an understanding confidant who may be able to help analyze and meet your problems. Very often, talking over your problems with someone in whom you have confidence and trust will help you to arrive at a solution.

3. Do not betray yourself by trying to evade the problem or by pretending that it has been solved. If social problem distractions prevent you from studying or doing satisfactory work, it is better to admit this to yourself. You can then decide what can be done about it.

Once you are free of interferences and irritations, it is much easier to stay focused on your studies.

Concentrate

To study effectively, you must concentrate. Your ability to concentrate is governed, to a great extent, by your surroundings as well as your physical condition. When absorbed in study, you must be oblivious to everything else around you. As you learn to concentrate and study, you must also learn to overcome all distractions. There are three kinds of distractions you may face:

1. Distractions in the surrounding area, such as motion, noise, and the glare of lights. The sun shining through a window on your study area, for example, can be very distracting.

   Some technicians find that, for effective study, it is necessary to eliminate visual distractions as well as noises. Others find that they are able to tolerate moderate levels of auditory or visual distraction.

   Make sure your study area is properly lighted and ventilated. The lighting should be adequate but should not shine directly into your eyes or be visible out of the corner of your eye. Also, try to avoid a reflection of the lighting on the pages of your book.

   Whether heated or cooled, the environment should be at a comfortable level. For most, this means a temperature of 78°F–80°F (25.6°C–26.7°C) with a relative humidity of 45 to 50 percent.

2. Distractions arising from your body, such as a headache, fatigue, and hunger. Be in good physical condition. Eat wholesome meals at regular times. Try to eat with your family or friends whenever possible. Mealtime should be your recreational period. Do not eat a heavy meal for lunch, and do not resume studies immediately after eating lunch. Just after lunch, try to get some regular exercise, relaxation, and recreation. A little exercise on a regular basis is much more valuable than a lot of exercise only on occasion.

3. Distractions of irrelevant ideas, such as how to repair the garden gate, when you are studying for an automotive-related test.

The problems associated with study are no small matter. These problems of distractions are generally best dealt with by a process of elimination. A few important rules for eliminating distractions follow.

Get Sufficient Sleep

You must get plenty of rest even if it means dropping certain outside activities. Avoid cutting in on your sleep time; you will be rewarded in the long run. If you experience difficulty going to sleep, do something to take your mind off your work and try to relax before going to bed. Some suggestions that may help include a little reading, a warm bath, a short walk, a conversation with a friend, or writing that overdue letter to a distant relative. If sleeplessness is an ongoing problem, consult a physician. Do not try any of the sleep remedies on the market, particularly if you are on medication, without approval of your physician.

If you still have difficulty studying, a final rule may help. Sit down in a favorable place for studying, open your books, and take out your pencil and paper. In a word, go through the motions.

Arrange Your Area

Arrange your chair and work area. To avoid strain and fatigue, whenever possible, shift your position occasionally. Try to be comfortable; however, avoid being too comfortable. It is nearly impossible to study rigorously when settled back in a large easy chair or reclining leisurely on a sofa.

When studying, it is essential to have a plan of action, a time to work, a time to study, and a time for pleasure. If you schedule your day and adhere to the schedule, you will eliminate most of your efforts and worries. A plan that is followed, then, soon becomes the easy and natural routine of the day. Most technicians find it useful to have a definite place and time to study. A particular table and chair should always be used for study and intellectual work. This place will then come to mean study. To be seated in that particular location at a regularly scheduled time will automatically lead you to assume a readiness for study.

Don't Daydream

Daydreaming or mind-wandering is an enemy of effective study. Daydreaming is frequently due to an inadequate understanding of words. Use the Glossary or a dictionary to look up the troublesome word. Another frequent cause of daydreaming is a deficient background in the present subject matter. When this is the problem, go back and review the subject matter to obtain the necessary foundation. Just one hour of concentrated study is equivalent to ten hours with frequent lapses of daydreaming. Be on guard against mind-wandering, and pull yourself back into focus on every occasion.

Study Regularly

A system of regularity in study is believed by many scholars to be the secret of success. The daily time schedule must, however, be determined on an individual basis. You must decide how many hours of each day you can devote to your studies. Few technicians really are aware of where their leisure time is spent. An accurate account of how your days are presently being spent is an important first step toward creating an effective daily schedule.

Weekly Schedule							
	Sun	Mon	Tues	Wed	Thu	Fri	Sat
6:00							
6:30							
7:00							
7:30							
8:00							
8:30							
9:00							
9:30							
10:00							
10:30							
11:00							
11:30							
NOON							
12:30							
1:00							
1:30							
2:00							
2:30							
3:00							
3:30							
4:00							
4:30							
5:00							
5:30							
6:00							
6:30							
7:00							
7:30							
8:00							
8:30							
9:00							
9:30							
10:00							
10:30							
11:00							
11:30							

The convenient form is for keeping an hourly record of your week's activities. If you fill in the schedule each evening before bedtime, you will soon gain some interesting and useful facts about yourself and your use of your time. If you think over the causes of wasted time, you can determine how you might better spend your time. A practical schedule can be set up by using the following steps.

1. Mark your fixed commitments, such as work, on your schedule. Be sure to include classes and clubs. Do you have sufficient time left? You can arrive at an estimate of the time you need for studying by counting the hours used during the present week. An often-used formula, if you are taking classes, is to multiply the number of hours you spend in class by two. This provides time for class studies. This is then added to your work hours. Do not forget time allocation for travel.

2. Fill in your schedule for meals and studying. Use as much time as you have available during the normal workday hours. Do not plan, for example, to do all of your studying between 11:00 P.M. and 1:00 A.M. Try to select a time for study that you can use every day without interruption. You may have to use two or perhaps three different study periods during the day.

3. List the things you need to do within a time period. A one-week time frame seems to work well for most technicians. The question you may ask yourself is: "What do I need to do to be able to walk into the test next week, or next month, prepared to pass?"

4. Break down each task into smaller tasks. The amount of time given to each area must also be settled. In what order will you tackle your schedule? It is best to plan the approximate time for your assignments and the order in which you will do them. In this way, you can avoid the difficulties of not knowing what to do first and of worrying about the other things you should be doing.

5. List your tasks in the empty spaces on your schedule. Keep some free time unscheduled so you can deal with any unexpected events, such as a dental appointment. You will then have a tentative schedule for the following week. It should be flexible enough to allow some units to be rearranged if necessary. Your schedule should allow time off from your studies. Some use the promise of a planned recreational period as a reward for motivating faithfulness to a schedule. You will more likely lose control of your schedule if it is packed too tightly.

Keep a Record

Keep a record of what you actually do. Use the knowledge you gain by keeping a record of what you are actually doing so you can create or modify a schedule for the following week. Be sure to give yourself credit for movement toward your goals and objectives. If you find that you cannot study productively at a particular hour, modify your schedule so as to correct that problem.

Scoring the ASE Test

You can gain a better perspective about tests if you know and understand how they are scored. ASE's tests are scored by American College Testing (ACT), a nonpartial, nonbiased organization having no vested interest in ASE or in the automotive industry. Each question carries the same weight as any other question. For example, if there are fifty questions, each is worth 2 percent of the total score. The passing grade is 70 percent. That means you must correctly answer thirty-five of the fifty questions to pass the test.

Understand the Test Results

The test results can tell you:

- where your knowledge equals or exceeds that needed for competent performance, or
- where you might need more preparation.

The test results *cannot* tell you:

- how you compare with other technicians, or
- how many questions you answered correctly.

Your ASE test score report will show the number of correct answers you got in each of the content areas. These numbers provide information about your performance in each area of the test. However, because there may be a different number of questions in each area of the test, a high percentage of correct answers in an area with few questions may not offset a low percentage in an area with many questions.

It may be noted that one does not "fail" an ASE test. The technician who does not pass is simply told "More Preparation Needed." Though large differences in percentages may indicate problem areas, it is important to consider how many questions were asked in each area. Since each test evaluates all phases of the work involved in a service specialty, you should be prepared in each area. A low score in one area could keep you from passing an entire test.

Note that a typical test will contain the number of questions indicated above each content area's description. For example:

Electrical/Electronic Systems (Test T6)

Content Area	Questions	Percent of Test
A. General Electrical System Diagnosis	11	22%
B. Battery Diagnosis and Repair	6	12%
C. Starting System Diagnosis and Repair	8	16%
D. Charging System Diagnosis and Repair	8	16%
E. Lighting Systems Diagnosis and Repair	6	12%
1. Headlights, Daytime Running Lights, Parking, Clearance, Tail, Cab, and Dash Lights (3)		
2. Stoplights, Turn Signals, Hazard Lights, and Backup Lights (3)		
F. Gauges and Warning Devices Diagnosis and Repair	6	12%
G. Related Systems	5	10%
Total	*50	100%

*__Note:__ *The test could contain up to ten additional questions that are included for statistical research purposes only. Your answers to these questions will not affect your score, but since you do not know which ones they are, you should answer all questions in the test. The five-year Recertification Test will cover the same content areas as those listed above. However, the number of questions in each content area of the Recertification Test will be reduced by about one-half.*

"Average"

There is no such thing as average. You cannot determine your overall test score by adding the percentages given for each task area and dividing by the number of areas. It doesn't work that way because there generally are not the same number of questions in each task area. A task area with twenty questions, for example, counts more toward your total score than a task area with ten questions.

So, How Did You Do?

Your test report should give you a good picture of your results and a better understanding of your task areas of strength and weakness.

If you fail to pass the test, you may take it again at any time it is scheduled to be administered. You are the only one who will receive your test score. Test scores will not be given over the telephone by ASE nor will they be released to anyone without your written permission.

Pretest

The purpose of this pretest is to determine the amount of review that you may require prior to taking the ASE medium/heavy truck test: Electrical/Electronic Systems (Test T6). If you answer all of the pretest questions correctly, complete the sample test in section 5 along with the additional test questions in section 6.

If two or more of your answers to the pretest questions are wrong, study section 4, An Overview of the System, before continuing with the sample test and additional test questions.

The pretest answers and explanations are located at the end of the pretest.

1. Which of these is most likely to be done when checking a circuit for continuity?
    A. Disconnecting the battery
    B. Connecting a voltmeter in series
    C. Turning the ignition on
    X D. Using an ohmmeter

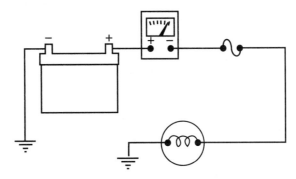

2. The meter in the figure shown above is being used to check:
    X A. current.
    B. resistance.
    C. voltage.
    D. watts.

3. A charging system check shows a low voltage output. Technician A says a defective voltage regulator could cause this. Technician B says an improperly adjusted alternator belt could cause this. Who is right?
    A. Technician A only
    B. Technician B only
    C. Both A and B
    D. Neither A nor B

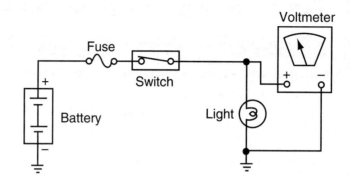

4. The switch is closed in the figure shown above and the battery is fully charged. The voltmeter indicates that the voltage drop across the light bulb is 9 volts. Technician A says that there may be a short to ground between the switch and the light. Technician B says the resistance in the circuit may be too high, causing low voltage at the bulb. Who is right?
   A. Technician A only
   B. Technician B only
   C. Both A and B
   D. Neither A nor B

5. When checking voltage drop on a starter motor, which of these procedures is the LEAST likely to be performed?
   A. Make sure the battery is fully charged
   B. Check the voltage drop while cranking the engine
   C. Check the voltage drop across both battery cables
   D. Disconnect the engine ground cable

6. A horn operates continuously without pressing the horn button. Which of these would be the most likely cause?
   A. An open wire to the relay
   B. A shorted horn coil
   C. The horn has a poor ground
   D. A shorted horn relay

7. When performing a resistance check with an ohmmeter, technician A says you should use a high impedance (10 megohm input) ohmmeter when checking sensitive electronic components. Technician B says an ohmmeter should be connected to a circuit in which the current is flowing. Who is right?
   A. Technician A only
   B. Technician B only
   C. Both A and B
   D. Neither A nor B

8. The battery is disconnected on a vehicle with on-board computers. This procedure may cause:
   A. failure of the engine to start after the battery is reconnected.
   B. a voltage surge in the electrical system.
   C. erasure of the computer electronically retained memory or RAM.
   D. damage to the on-board computers.

9. All of these statements about replacement of a halogen headlamp bulb are true **EXCEPT:**
   A. keep the bulb free of moisture or contaminants.
   B. change the bulb with the headlights on.
   C. handle the bulb only by its base.
   D. do not drop or scratch the bulb.

10. When using a 12-volt test light to test for voltage, all of the following are true **EXCEPT:**
    A. the test lamp may be connected to the chassis ground.
    B. the battery ground should be disconnected.
    C. the test lamp may be connected to the battery ground.
    D. the ignition can be turned on or off.

11. A truck with an erratic fuel gauge is brought in. What should be done first?
    A. Check battery voltage
    B. Remove fuel gauge and send out for repair
    C. Check sending unit ground connection
    D. Verify the customer's complaint

12. While bench testing a rotor for continuity, a technician reads 1 ohm across the slip rings with an ohmmeter. Which of these indicates the LEAST likely winding condition?
    A. Shorted
    B. Grounded
    C. Normal
    D. Burned

Answers to the Test Questions for the Pretest

1. D, 2. A, 3. C, 4. B, 5. D, 6. D, 7. A, 8. C, 9. B, 10. B, 11. D, 12. C

Explanations to the Answers for the Pretest

Question #1
Answer A is wrong. The technician would leave the battery connected when using an unpowered test lamp or a voltmeter for a continuity check.
Answer B is wrong. An ammeter would be connected to check current flow. Voltmeters are connected across the circuit for voltage or voltage drops.
Answer C is wrong. The ignition does not necessarily have to be on for a continuity check unless the circuit is in the series with the ignition switch.
Answer D is correct because in most cases one uses an ohmmeter to check continuity.

Question #2
Answer A is correct. Measuring a circuit in series requires an ammeter.
Answer B is wrong. An ohmmeter may be used to check resistance.
Answer C is wrong. Voltage is checked across the circuit, not in a series as indicated in the figure.
Answer D is wrong. Watts is a measurement that is calculated by volts × amps.

Question #3

Answer A is wrong because there is more than one correct answer and both technicians are correct. A faulty voltage regulator can cause a low voltage output and low battery.
Answer B is wrong because there is more than one correct answer and both technicians are correct. An improperly adjusted belt can cause the alternator pulley to slip and cause a low voltage output.
Answer C is correct. A faulty voltage regulator may cause a low voltage output, and thereby, a low battery. In addition, an improperly adjusted belt may cause the alternator pulley to slip and cause a low voltage output from the alternator.
Answer D is wrong.

Question #4

Answer A is wrong because a short to ground between the switch and bulb will cause the fuse to blow.
Answer B is correct. If the voltmeter shows a voltage drop less than battery voltage then there is higher than normal resistance in the circuit.
Answer C is wrong because one technician is wrong.
Answer D is wrong because technician B is right.

Question #5

Answer A is wrong. The battery must be at full charge before testing.
Answer B is wrong. You also check the circuit voltage drops dynamically with the engine cranking.
Answer C is wrong because you also need to check both the positive and negative cables.
Answer D is correct. It is not necessary to disconnect the engine ground during these tests.

Question #6

Answer A is wrong because an open wire would not allow the horn to operate continuously.
Answer B is wrong because a shorted horn coil causes the horn not to sound at all.
Answer C is wrong because a poor ground allows very little, if any, horn operation.
Answer D is correct because a shorted horn relay can allow the horn to operate continuously.

Question #7

Answer A is correct because using a high impedance ohmmeter is less likely to damage sensitive electronic components.
Answer B is wrong. When using an ohmmeter on a circuit, the power should be turned off or damage to the ohmmeter will result.
Answer C is wrong because only one technician is right.
Answer D is wrong because technician A is right.

Question #8

Answer A is wrong. Just disconnecting and reconnecting the battery will not affect the starting of the vehicle.
Answer B is wrong. Disconnecting the battery will not cause a voltage surge.
Answer C is correct. Most automotive computer controllers have a volatile memory that is erased when the power is disconnected.
Answer D is wrong. Disconnecting the battery will not damage the on-board computers.

Question #9
Answer A is wrong. A halogen bulb should be kept free of moisture and contaminants.
Answer B is correct because a technician should never change a halogen bulb or any other electrical component when there is power in the circuit.
Answer C is wrong. A halogen bulb should be handled by its base to prevent oil from the skin contaminating the glass envelope.
Answer D is wrong. A fragile object should be handled carefully.

Question #10
Answer A is wrong because the technician can connect the test lamp to any viable ground on the vehicle.
Answer B is correct because you do not have to disconnect the battery ground when using a 12-volt test light.
Answer C is wrong because the technician can connect the test lamp to any viable ground on the vehicle.
Answer D is wrong because the ignition can be on or off depending on the nature of the circuit.

Question #11
Answer A is wrong because a voltage check is part of the process after verifying the customer's complaint.
Answer B is wrong because removal of the gauge assembly may be the last step in this repair process.
Answer C is wrong because a ground check is performed after the voltage check.
Answer D is correct because the technician must first properly verify the customer's complaint.

Question #12
Answer A is wrong because in this case a shorted rotor field coil at less than specification is the most likely cause, not the least.
Answer B is wrong because the procedure for a ground check is connecting the ohmmeter from one slip ring to the rotor poles.
Answer C is correct because normal is the least likely condition in this case. Measuring 1 ohm of resistance across the slip rings indicates a shorted rotor, which is abnormal.
Answer D is wrong because a burned rotor is a visual check, not one requiring a continuity check.

Types of Questions

ASE certification tests are often thought of as being tricky. They may seem to be tricky if you do not completely understand what is being asked. The following examples will help you recognize certain types of ASE questions and avoid common errors.

Each test is made up of forty to eighty multiple-choice questions. Multiple-choice questions are an efficient way to test knowledge. To answer them correctly, you must think about each choice as a possibility, and then choose the one that best answers the question. To do this, read each word of the question carefully. Do not assume you know what the question is about until you have finished reading it.

Multiple-Choice Questions

One type of multiple-choice question has three wrong answers and one correct answer. The wrong answers, however, may be almost correct, so be careful not to jump at the first answer that seems to be correct. If all the answers seem to be correct, choose the answer that is the most correct. If you readily know the answer, this kind of question does not present a problem. If you are unsure of the answer, analyze the question and the answers. For example:

Question 1:

A stator's wires are grouped into three separate bundles, also known as:
A. diodes.
B. windings.
C. rectifiers.
D. poles.

Analysis:

Answer A is wrong because diodes are electrical check valves mounted in the rectifiers. **Answer B is correct** because the stator contains the windings that become the electrical conductor in the charging circuit.

Answer C is wrong because rectifiers contain diodes and are used to convert AC to DC. Answer D is wrong because poles is a term for electrical tereminals.

EXCEPT Questions

Another type of question used on ASE tests has answers that are all correct except one. The correct answer for this type of question is the answer that is wrong. The word **EXCEPT** will always be in capital letters. You must identify which of the choices is the wrong answer. If you read quickly through the question, you may overlook what the question is asking and answer the question with the first correct statement. This will make your answer wrong. An example of this type of question and the analysis is as follows:

Question 2:

A truck has an intermittent fault with its high beam headlamps. All of these could be a possible cause **EXCEPT:**
A. a defective headlight dimmer switch.
B. a defective high beam filament inside the headlamps.
C. a loose wiring harness connector.
D. a defective headlight switch.

Analysis:

Answer A is wrong because a defective headlight dimmer switch could cause a faulty high beam operation.

Answer B is wrong because a defective high beam filament could cause partial or total failure of the high beam circuit.

Answer C is wrong because loose connectors to the high beam circuit will cause intermittent problems.

Answer D is correct because a defective headlight switch could not be the cause of an intermittent headlight problem.

Technician A, Technician B Questions

The type of question that is most popularly associated with an ASE test is the "Technician A says . . . Technician B says . . . Who is right?" type. In this type of question, you must identify the correct statement or statements. To answer this type of question correctly, you must carefully read each technician's statement and judge it on its own merit to determine if the statement is true.

Typically, this type of question begins with a statement about some analysis or repair procedure. This is followed by two statements about the cause of the problem, proper inspection, identification, or repair choices. You are asked whether the first statement, the second statement, both statements, or neither statement is correct. Analyzing this type of question is a little easier than the other types because there are only two ideas to consider although there are still four choices for an answer.

Technician A . . . Technician B questions are really double-true-false questions. The best way to analyze this kind of question is to consider each technician's statement separately. Ask yourself, is A true or false? Is B true or false? Then select your answer from the four choices. An important point to remember is that an ASE Technician A . . . Technician B question will never have Technician A and B directly disagreeing with each other. That is why you must evaluate each statement independently. An example of this type of question and the analysis of it follows.

Question 3:

A truck has a turn signal complaint. You find that the left front turn light does not flash and the left rear light flashes faster than normal. Technician A says that the left front bulb could be defective. Technician B says there could be an open circuit between the switch and the left front bulb. Who is right?

A. A only
B. B only
C. Both A and B
D. Neither A nor B

Analysis:

Answer A is wrong. It is a good choice because if the left front bulb cannot flash, the load changes on the flasher and the flashers will flash at a high frequency. Yet, this is the wrong answer because both technicians are correct.

Answer B is wrong. It is a good choice because an open also changes the load on the flasher and the flashers will flash at a high frequency plus the light will not flash or light. Yet, this is the wrong answer because both technicians are correct.

Answer C is correct.

Answer D is wrong.

Questions with a Figure

About 10 percent of ASE questions will have a figure, as shown in the following example:

Question 4:

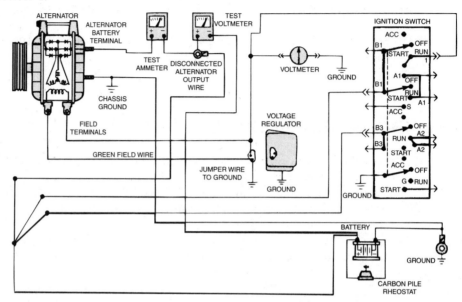

With an ammeter and voltmeter connected to the charging system, as shown in the figure above, the voltmeter indicates 2 volts and the ammeter reads 10 amps. Technician A says this condition may cause an undercharged battery. Technician B says this condition may result in a headlight flare-up during acceleration. Who is right?

A. A only
B. B only
C. Both A and B
D. Neither A nor B

Analysis:

Answer A is correct because the technician is performing a voltage drop test from the alternator battery wire to the battery positive cable. A 2-volt drop indicates excessive resistance and will reduce charging current.

Answer B is wrong because this condition will cause dim headlights.

Answer C is wrong because technician B is wrong.

Answer D is wrong because technician A is correct.

Most-Likely Questions

Most-likely questions are somewhat difficult because only one choice is correct while the other three choices are nearly correct. An example of a most-likely-cause question is as follows:

Question 5:

A starter control circuit test will most likely identify which of the following conditions?

A. High resistance in the solenoid switch circuit
B. A loose neutral safety switch
C. A loose battery cable condition
D. Short in the starter armature windings

Analysis:

Answer A is correct because high resistance in the starter solenoid switch circuit would be located during a control circuit check.

Answer B is wrong because the neutral safety switch is not considered part of the control circuit.

Answer C is wrong because the battery cables are not considered part of the control circuit.

Answer D is wrong because the starter armature is not considered part of the control circuit.

LEAST-Likely Questions

Notice that in most-likely questions there is no capitalization. This is not so with least-likely type questions. For this type of question, look for the choice that would be the least likely cause of the described situation. Read the entire question carefully before choosing your answer. An example is as follows:

Question 6:

What is the LEAST likely result of a full-fielded alternator?
A. High battery voltage level
B. Battery gassing
C. Low battery voltage level
D. High alternator amperage output

Analysis:

Answer A is wrong. A very high battery voltage can be expected because a full battery voltage is being sent directly to the field windings.

Answer B is wrong because battery gassing is possible on a full-fielded alternator due to very high system voltages.

Answer C is correct because full-fielding will most likely generate a high voltage level.

Answer D is wrong. High alternator amperage will be generated when full-fielding occurs because unregulated current is being supplied to the field windings.

Summary

There are no four-part multiple-choice ASE questions having "none of the above" or "all of the above" choices. ASE does not use other types of questions, such as fill-in-the-blank, completion, true-false, word-matching, or essay. ASE does not require you to draw diagrams or sketches. If a formula or chart is required to answer a question, it is provided for you. There are no ASE questions that require you to use a pocket calculator.

Testing Time Length

An ASE test session is four hours and fifteen minutes. You may attempt from one to a maximum of four tests in one session. It is recommended, however, that no more than a total of 225 questions be attempted at any test session. This will allow for just over one minute for each question.

Visitors are not permitted at any time. If you wish to leave the test room, for any reason, you must first ask permission. If you finish your test early and wish to leave, you are permitted to do so only during specified dismissal periods.

Monitor Your Progress

You should monitor your progress and set an arbitrary limit to how much time you will need for each question. This should be based on the number of questions you are attempting. It is suggested that you wear a watch because some facilities may not have a clock visible to all areas of the room.

Registration

Test centers are assigned on a first-come, first-served basis. To register for an ASE certification test, you should enroll at least six weeks before the scheduled test date. This should provide sufficient time to assure you a spot in the test center. It should also give you enough time for study in preparation for the test. Test sessions are offered by ASE twice each year, in May and November, at over six hundred sites across the United States. Some tests that relate to emission testing also are given in August in several states.

To register, contact Automotive Service Excellence/American College Testing at:

ASE/ACT
P.O. Box 4007
Iowa City, IA 52243

4 An Overview of the System

Electrical/Electronic Systems (Test T6)

The following section includes the task areas and task lists for this test and a written overview of the topics covered in the test.

The task list describes the actual work you should be able to do as a technician that you will be tested on by the ASE. This is your key to the test and you should review this section carefully. We have based our sample test and additional questions upon these tasks, and the overview section will also support your understanding of the task list. ASE advises that the questions on the test may not equal the number of tasks listed; the task lists tell you what ASE expects you to know how to do and be ready to be tested on.

At the end of each question in the Sample Test and Additional Test Questions sections, a letter and number will be used as a reference back to this section for additional study. Note the following example: **A1.**

Task List

A. General Electrical System Diagnosis (11 Questions)

Task A1   Check continuity in electrical/electronic circuits using appropriate test equipment.

Example:

1. Which of these is most likely to be done when checking a circuit for continuity?
    A. Connecting an ammeter in parallel
    B. Connecting a voltmeter in series
    C. Turning the ignition on
    D. Using an ohmmeter   (A1)

Question #1

Answer A is wrong. An ammeter is not used to check continuity.

Answer B is wrong. An ammeter would be connected to check current flow. Voltmeters are connected across the circuit for voltage or voltage drops.

Answer C is wrong. The ignition does not necessarily have to be on for a continuity check unless the circuit is in the series with the ignition switch.

Answer D is correct because in most cases one uses an ohmmeter to check continuity.

Task List and Overview

A. General Electrical System Diagnosis (11 Questions)

Task A1   **Check continuity in electrical/electronic circuits using appropriate test equipment.**

Checking continuity in an electrical circuit is one of the most common troubleshooting tests a technician will make. While the purpose of this test is to make sure that a complete current path exists in the circuit being tested, it is important to remember that it is not an accurate indication of overall circuit "health" (e.g., excessive resistance). A continuity test is most useful to quickly differentiate one circuit from another, such as trying to locate a specific contact in a multiple pin connector.

Continuity tests can be made a number of different ways. If the circuit is under power, a 12/24-volt test light or voltmeter can be used to check for voltage at various test points in a particular circuit. If the circuit is not energized and disconnected at both ends (the most common way to check continuity), then a self-powered test light or ohmmeter can be used to check for a complete circuit. Both tools accomplish this by energizing the circuit with low potential current to determine that the circuit can be closed. Many DMMs (digital multimeters) have a separate feature on them that will allow continuity tests to be made simply by listening for an audible beep. This is handy because multiple tests can be made rather quickly without having to look constantly at the display for a resistance value.

When testing circuits that include ECMs (electronic control modules), it is essential to observe the OEM diagnostic procedure to avoid possible damage to the processor.

Task A2   **Check applied voltages, circuit voltages, and voltage drops in electrical/electronic circuits using a digital multimeter (DMM).**

Voltage tests are always measured with the test leads of the DMM in parallel to the component or circuit being tested. In testing applied voltage, the negative or black test lead of the DMM is connected to a battery or chassis ground. The positive red lead is then probed near the power source (e.g., a switch or fuse) to determine if the circuit is receiving the proper voltage. A handy feature of a DMM is that polarity is not important. If the leads are reversed, the display will simply include a minus sign in front of the reading. Circuit voltages are tested much the same way. Simply take the positive test probe and go from one circuit's power source to the next. Generally, the applied voltages should all be the same.

When selecting a DMM to use for testing electronic circuits, it is recommended that it have a 10 megohm or higher impedance. This is necessary to limit the affect the unit might have on accuracy when testing low current flow circuits. For this same reason, never use a test light to check for voltage in an electronic circuit. The lamp in the test light will draw too much current and this can damage the circuit integrity.

If you need to test for voltage in an energized circuit that is in operation, a convenient way to do this is with a back probe tool, also known as a "spoon," at various connectors in a particular circuit. This DMM accessory allows you to probe into the connector from its backside without disconnecting it.

Voltage drops are often misunderstood. A voltage drop is simply the loss of voltage (electrical "pressure") in a circuit due to unwanted resistance. Remember, that for there to be a voltage drop, there has to be current flow through the circuit being tested. Any resistance other than the designated load (e.g., light bulb, blower motor) will produce unwanted voltage drop along with a reduced current flow in that circuit. As with the voltage test, the test leads are probed in parallel to the component or portion of the circuit being tested. If, for example, two mating contacts in a joined connector assembly were suspect, then both sides of the connector would be probed with the test leads. Any voltage reading with the circuit under load is a voltage drop. As a rule of thumb, each connection, switch, fuse, length of wire, etc., should see no more than one-tenth of a volt (100 millivolts) drop while under load.

Task A3   Check current flow in electrical/electronic circuits and components using an ammeter, digital multimeter (DMM), or clamp-on ammeter.

Current flow tests are used when one suspects a circuit that has a higher than normal current flow, such as a dragging blower motor, or a circuit with a high-resistance short to ground. In other words, situations that might continually blow a fuse or trip a circuit breaker.

In testing current flow with an ammeter or DMM, it is important to remember that the test leads are connected in series with the circuit being tested, usually at a point near the power source. The circuit must be broken at some point to allow the connection of the test leads.

Most DMMs have a 10 to 20 amp limit when measuring amperage directly through the meter. Any greater current flow will blow its fuse. If you suspect that the circuit carries more than that, then a safe way to test for current flow would be to use a current clamp. This device simply clamps over the wire being tested and determines current flow by measuring the strength of the magnetic field surrounding the wire. While it is extremely handy to use, it is not as accurate as routing all the current through the DMM, especially in circuits flowing less than 10 amps.

Remember: Low current flow usually is a result of excessive resistance in a circuit or low voltage. Higher than normal current flow can generally be traced to excessive applied voltage or a shorted component or wire.

Task A4   Check resistance in electrical/electronic circuits and components using an ohmmeter or digital multimeter (DMM).

Resistance checks are typically made when a circuit has unwanted voltage drops or low current flow. An ohmmeter is a device that circulates a small current through a circuit when it is not energized and then measures the voltage drop through it. It displays this resistance (or restriction to current flow) in units known as ohms. The lower the ohmic value, the less restriction to electron flow there will be in a circuit. If the meter reads infinity (or a high flashing number on most DMMs), this means that the circuit is open. Except for where a resistance is built into a circuit, such as a blower motor resistor or a load itself, generally speaking the lower resistance a circuit has, the better. For example, when testing a length of wire or a fuse, most will test very near zero ohms. When testing with a DMM, always be sure to zero the meter first to compensate for any resistance present in the test leads, especially when testing low resistance components or circuits.

When making tests with a DMM that is not auto-ranging, be sure to select a range that will give accuracy. For example, if you are testing a component with an expected 10,000-ohm reading on, set your scale to "×1000" (multiply by 1000) so that the reading on your meter will be around 10. Similarly, testing a component with two ohms resistance on the "×1000" scale will not be accurate. In this case, set the meter to "×1." If you are using an analog meter, set the meter to a range setting that will put the needle in roughly the middle of the scale for the component you are testing.

Resistance checking of specific components is generally reserved for when a manufacturer specifies a certain test value, such as a fuel level sending unit. Some components, such as light bulbs and glow plugs, do not lend themselves to resistance testing because their resistance changes dramatically as they heat up. Also, large diameter conductors such as battery cables cannot be reliably tested with an ohmmeter because it cannot circulate enough current to simulate actual operating conditions and identify a restriction. Testing of battery cables is best done using a voltage drop test (see Task C1).

Task A5   Find shorts, grounds, and opens in electrical/electronic circuits.

A short circuit is defined as one where the current flow is allowed to escape to ground at a point other than where it was designed, such as a bare wire rubbing against the frame. An open circuit is usually caused by a broken wire or other component not making the necessary connection to complete a circuit, stopping current flow in that portion of the circuit.

Finding shorts is best done with an ohmmeter because with a live circuit a fuse or breaker will continually blow. Do not install a larger fuse because there will be a risk of melting a bundle of wires. Divide the circuit apart into smaller lengths at various connectors (where applicable) while testing for continuity to ground (there should be none), or visually inspect the harness for rub or pinch points.

Locating opens can sometimes be more difficult, because sometimes the damage is not visually apparent. A good way to test for an open would be to apply voltage at one end and then probe at each succeeding point or connection downstream until you find no power. Alternatively, an ohmmeter can be used to do the same thing when the circuit is not energized.

It is important to remember when testing at connectors, especially smaller ones designed for electronic circuits, that you do not damage the contacts when probing into them, especially female ones. Spreading apart the tangs on a female contact while testing will create more problems than you began with. Always use an appropriate adapter for making tests to these types of contacts.

Task A6   Diagnose key-off battery drain problems.

While dead batteries are not always caused by a key-off battery drain problem, when you suspect this it is a good idea before you start troubleshooting to know the possible causes and how the system is constructed. Before the the widespread use of electronics in vehicles, many systems had zero current draw with the key off. Today, with so many vehicles having one or more processors, testing for battery drain will likely involve unplugging connectors from ECMs or isolating circuits.

In older vehicles, disconnect the negative battery cable and hook up your ammeter in series with it and the ground post. Make sure the key is off and all loads such as dome lights are turned off. There should be no current draw. If there is, an easy way to isolate the problem is to start pulling fuses one at a time until the draw stops. Another possible draw is through a defective diode in the alternator. Disconnect the positive lead at the alternator to locate this potential problem.

With electronic control modules, it is best to have the manufacturer's specifications to be certain that a draw is within parameters (most will draw well under 50 milliamps with the key off). If it measures higher than specified, it might be necessary to disconnect its power supply to be sure that there is not another component in the vehicle causing the additional draw. Be sure to allow time for the ECM to power down and enter the "sleep mode" before taking a final reading.

Also, when checking for a battery draw complaint, check surface discharge across the top of the battery (see Task B3).

Task A7   Inspect, test, and replace fusible links, circuit breakers, and fuses.

A fuse is an electrical safety device. When it blows, it is because of a current overload somewhere in a circuit. Always repair the problem; never install a fuse of a higher rating. Also, learn to identify the reason for a fuse failure. If the metal link in the center of the fuse melted, it was caused by a current overload. On older glass style fuses, if the end caps are found to be melted, this was caused by poor or corroded contacts in the fuse holder itself, not a current overload.

A circuit breaker performs the same function as a fuse; however, it has a feature that will allow it to be reset after it trips, usually automatically. Most circuit breakers can be identified as a small rectangular box with two studs attached to it. They will also have their maximum current rating stamped on it. Two types of circuit breaker are used in truck electrical circuits. SAE #1 circuit breakers cycle when overloaded. SAE #2 circuit breakers trip when overloaded and do not reset until the circuit is opened. Circuit breakers are handy test devices to have in a tool box. When testing a circuit that continually blows conventional fuses, installing a circuit breaker into the circuit temporarily with jumper wires saves both time and money.

Fusible links are special short pieces of wire designed to melt in half in case of overload. They are usually installed near a power source (e.g., battery or starter solenoid) and are normally two to four wire gauge sizes smaller in diameter than the circuit they are protecting. When they do melt in half, the insulation usually bubbles, but not always, making them difficult to troubleshoot. The fuse link has a special high-temperature insulation designed not to separate during an overload. Sometimes the easiest way to test these devices is to simply give them a good tug at either end. If it stretches, it is bad.

Many manufacturers are now using what are known as maxi-fuses in place of fusible links, which are also usually located near the battery or main power distribution bus. These are simply conventional looking fuses much larger than standard. They are also much easier to install and troubleshoot.

To test any of the previously mentioned circuit safety devices, simply remove it and test for resistance across it with an ohmmeter. A good component should show very low resistance.

Task A8   Inspect, test, and replace spike suppression diodes/resistors and capacitors.

A diode is simply an electrical "check valve" that will allow current to flow in a circuit in only one direction. A symbol of a diode looks like an arrow with a line drawn perpendicular to its point. A typical use of a diode is in an alternator, where they perform the task of converting AC voltage into DC by simply preventing output to the battery during the negative portion of the sine wave. Diodes can also be used to prevent back feeding of current from the alternator excitation circuit to key switch circuits once the engine starts.

Testing diodes are a simple matter. You need only to take an ohmmeter and check for resistance (or continuity) through both sides, and in both directions. If it is good, current will flow in one direction but not the other.

Spike suppression devices can be in the form of diodes, resistors, or capacitors. Their purpose is to absorb or redirect a voltage spike that might come from any collapsing magnetic field, such as when the AC compressor clutch coil is switched off. By installing such a device in parallel with the coil, the voltage spike is simply absorbed back into the clutch coil and prevented from damaging potentially sensitive components, such as ECMs.

Task A9   Inspect, test, and replace relays and solenoids.

A relay is defined as a switching device that uses a small amount of current to control a larger one. A solenoid by definition is a device that performs mechanical movement on energizing, creating a magnetic field around an iron core, such as a fuel shutoff solenoid. A solenoid can also incorporate a relay function. A good example of this would be a starter solenoid, which not only moves the starter drive pinion into mesh with the flywheel ring gear, but also makes the high current connection between the battery and the starter field coils.

Most mini-relays have four or five terminals. The two small terminals (sometimes marked as #85 and #86) are used to energize the coil that creates the magnetic attraction necessary to cause a connection between the high amperage switch contacts. Two other terminals are needed to make the high amperage power in (terminal #30) and power out (terminal #87). Sometimes a fifth terminal (marked #87a) is used as a normally closed contact, as opposed to #87 being normally open, which closes when the relay energizes. Also note that the physical size of the power terminals (#30, #87, and #87a) may or may not be larger than the control terminals (#85 and #86) depending on the amperage capacity of the relay.

In operation, when a relay is signaled to close the high-current contacts, a small amount of current is fed through terminals #86 and #85, one being battery positive, and the other battery ground. This signal usually comes from an ECM, key switch, or other low-current switching device. The positive side of the high-current contact (#30) is then mated to the load side (#87), which completes the battery circuit to the high-current device, such as a horn or multiple light circuits. Some relays do not have the same code numbers, but work using exactly the same principles.

B. Battery Diagnosis and Repair (6 Questions)

Task B1   **Perform battery load test; determine needed service.**

To properly perform a battery load test, first determine whether or not the battery is fully charged. There is no point in testing a partially charged battery caused by charging system problems because it will fail. First determine the battery state of charge (see Task B2).

When you are sure the battery is ready to be load tested, first draw off the surface charge if the battery has just been charged (by either an alternator or battery charger). Load the battery by either cranking the starter for 15 seconds or drawing roughly 300 amps with the load tester for the same time period. Allow the battery to sit for a few minutes.

To perform the load test, first determine the rating of the battery. This is usually expressed in CCA (cold cranking amps), although some older batteries might be rated in AH (ampere-hours). Draw the battery down with the load tester at a rate equal to one-half the CCA or 3 times the AH. Hold this load for 15 seconds. At the end of 15 seconds, with the load still applied, note the battery voltage. Any reading over 9.6 volts at 70°F means the battery has passed the test. A battery that passes at 10.7 volts versus one that passes at 9.7 volts is the better of the two. One that barely passes will not be a long life battery, especially if it is much more than 5 or 6 years old. Keep this in mind in severe weather climates.

Task B2   **Determine battery state of charge by measuring terminal post voltage using a digital multimeter (DMM).**

A rough idea of the state of charge of a battery can be determined by measuring what is known as its open circuit voltage. With the battery out of the vehicle under no load, measure voltage across the top of the battery. If it is 12.6 volts or more, it is considered to be fully charged. As noted in Task B1, if the battery has been recently charged, either in the vehicle or out, draw off the surface charge using the method described earlier. Before testing, allow the battery to sit for 15 minutes, then proceed to test the open circuit voltage. Any reading less than 12.6 volts indicates that the battery should be charged.

Battery state of charge can also be determined using a hydrometer. A reading of 1.265 or higher at 80°F indicates a fully charged battery. A severely discharged battery will show around 1.120 or so. A reading of around 1.200 would indicate a battery that is 50 percent charged. Corrections must be made to the readings for batteries not within 10° of 80.

Task B3   **Inspect, clean, service, or replace battery and terminal connections.**

As part of any battery service routine, start by wearing eye protection. Inspect the top of the battery case for a buildup of dirt and moisture that can cause a low amperage current draw across the top of the battery. This can be checked by taking one probe of a DMM and dragging it across the top of the battery while holding the other probe on one of the posts. Any significant reading indicates a small current "short" between the two battery terminals, which can cause a low or dead battery over a period of time. A battery is best cleaned with a water and baking soda solution.

Battery cables and their terminal ends are a frequent source of problems. Many times they are the cause of a no-start or a sluggish starting complaint. A simple voltage drop test (see Task C1) will quickly identify which connection(s) have excessive resistance. Battery tapered type posts and cable ends are best cleaned with a scraper type tool that actually peels away all the old corrosion down to bright shiny metal. Flat, screw type posts are more difficult to clean; however, it is just as important that they be clean in order to transfer current with minimal voltage drop.

When reinstalling battery cable ends, coat the terminals with grease or petroleum jelly, or use a spray marketed for this purpose, to resist corrosion. Protective pads that go under the tapered terminals serve the same function.

When removing a battery and/or cables, always remove the negative cable first, and reconnect it last. This will help to prevent arcing and a possible battery explosion should your wrench come into contact with ground when loosening the positive cable.

If a conventional type battery is low on electrolyte level, add distilled water only. Never top up with acid.

Task B4   Inspect, clean, repair, or replace battery boxes, mounts, and holddowns.

The life of a battery depends in large part on the way it is secured in its mounts. A battery with missing straps or mounts will bounce around and eventually damage the internal separator plates, which may cause an internal short. Similarly, holddown straps tightened too tight may cause the case to crack.

Inspect the battery box when the battery is removed. Clean away any dirt, rust, and corrosion to help combat surface discharge when the battery is reinstalled.

Task B5   Charge battery using slow or fast charge method as appropriate.

The amount of time it takes to properly charge a battery depends on its state of charge. The other factor to consider is whether to charge it at a fast or slow rate. In most cases, a slow charge rate allows for maximum battery life and performance.

When slow charging, a 5–10 amp rate is usually sufficient, although a full charge may take overnight if the battery was quite low to start with. If time does not allow, a fast charge may be used; however, certain precautions must be followed. Never allow more than a 50–60 amp charge rate, and even then, closely monitor the electrolyte temperature to make sure it does not exceed 125°F. If the specific gravity reaches 1.225 during the fast charge, reduce the charge rate accordingly. If gassing or spewing of electrolyte occurs, reduce the charge rate. Never charge a frozen battery until it is brought up to room temperature.

Be sure to charge batteries in a well-ventilated area, away from sparks and other sources of heat. If the caps are removed during charging, cover the top of the battery with a moist rag. You can also monitor the specific gravity of the battery during charging to determine its state of charge. A reading of 1.265 indicates a full charge. Alternately, an ammeter on the charger that slowly drops to zero or near zero indicates a fully charged battery. Keep in mind though that a battery that will not accept a charge from the start (zero amps) is likely to be highly sulfated and should be discarded.

Task B6   Jump-start a vehicle using jumper cables and a booster battery or auxiliary power supply.

Jump-starting a vehicle with a dead battery is a simple procedure. However, it can be dangerous. The proper procedures must be followed to ensure that a spark is not generated that can cause an explosion. Keep in mind that a battery that is low due to prolonged cranking is likely generating a considerable amount of explosive vapors near the vent caps. A spark at this location could be dangerous.

Always wear eye protection when jump-starting a vehicle. Then, with both vehicle engines off, connect the positive terminals of both batteries with one cable. Connect the ground booster cable to the negative battery of the booster vehicle, and make the last connection to the frame or engine block of the dead vehicle. This will prevent sparks in the area of the dead battery.

Start the engine of the booster vehicle. If the jumper cables are of a generous size, crank and start the dead engine immediately. If the cables are small, or you are using a fast/boost charger, allow the connection to remain for a few minutes while the dead battery recharges. Then, with the help of the booster, crank and start the engine. Remove the cables in the reverse order of installation.

C. Starting System Diagnosis and Repair (8 Questions)

Task C1   **Perform starter circuit voltage drop tests; determine needed repairs.**

Cranking system related problems are one of the most common vehicle malfunctions that a technician will encounter. It is important to know how to perform a few simple tests in this area to quickly and effectively solve the problem. A cranking system voltage drop test identifies high resistance in the cranking circuit that can cause a slow or no-start condition. Every connection and conductor in the circuit, from the battery to the solenoid through the ground path is a potential problem area.

To test for voltage drop in the positive battery cable, set your DMM to the volts scale and attach one probe to the positive terminal post of the battery (not the cable or terminal). Touch the other probe of your DMM to the starter solenoid stud (again, not the cable itself). While holding the probes in this position, crank the engine until you get a steady reading. Ideally, there should be less than a half-volt drop. If there is considerably more, you can narrow the problem down by checking point to point along the cable (e.g., battery post to cable end, cable end to cable wire, wire end to wire end). Follow the same procedure noted earlier. (Polarity with a DMM is not important. If you have the leads backwards, the display will simply include a "minus" sign in front of the reading.) Ideally, each connection along the cable should have less than one-tenth of a volt drop.

Also, a quick way to check for voltage drop is to crank the engine for several seconds and then feel along all the connection points and conductors for heat. If any point is more than just warm, there is an excessive voltage drop at that location. Clean or replace as required.

Task C2   **Inspect, test, and replace components (key switch, push buttons, and/or magnetic switch) and wires in the starter control circuit.**

The starter control circuit includes those components between the key start switch and the magnetic switch (or starter solenoid itself if there is no magnetic switch in the circuit). It is important to remember that this is a critical, high-current circuit. Current requirements of solenoids on some larger starters are upwards of 100 amps, often necessitating a separate relay (also known as a magnetic switch) to handle these high amperage loads. Magnetic switch assemblies are typically 4 post units, having two large and two small terminals. However, some can also resemble large 5 pin mini-relays.

With this in mind, remember that the key start switch is a low-current switching device that controls the magnetic switch (relay). Current from this relay then goes to the starter solenoid (another relay) to engage the starter. So in reality we have several different circuits needed in working order to get the engine to crank. Also, between the key switch and the magnetic relay are sometimes found neutral safety switches, designed to prevent an engine from starting in gear.

Some manufacturers do not use a separate magnetic switch between the key switch and the starter solenoid, which makes the key switch circuit a fairly high-current one. Throw in a neutral safety switch and some bulkhead harness connections, and we have the potential for trouble in the form of a no-start condition if everything is not in good working order.

It is best to have the wiring diagram for the vehicle you're working on to see what components are used to effectively troubleshoot the starting control circuit. With your knowledge of relays (see Task A9) and voltage drop testing (see Task C1), diagnosing starting circuit complaints need not be intimidating.

Task C3   **Inspect, test, and replace starter relays and solenoids/switches.**

Most starter solenoids do two things: first, they shift the starter drive pinion into mesh with the flywheel ring gear, and at the same time they also make the high-current connection to deliver battery power to the starter motor field coils.

If a starter fails to crank, and all the other circuits to the solenoid check out, remember that we still have a set of electrical contacts inside the solenoid that are subject to pitting and corrosion over time. Since these are not easily seen, they often become overlooked during troubleshooting. However, testing them is easy. Perform a voltage drop test across the two large posts on the solenoid while cranking the engine, just as you would for a battery cable or clamp.

If a starter pinion fails to engage the ring gear, this could be due to a solenoid fault. When replacing a solenoid, keep in mind that quite often a pinion clearance adjustment needs to be made at the same time.

Task C4   **Remove and replace starter; inspect flywheel ring gear or flex plate.**

Whenever removing a starter motor, carefully inspect the teeth on both the starter drive pinion and the flywheel ring gear. Compare with new components if unsure of what is worn. Keep in mind that both the pinion and ring gear have a chamfer on the teeth to enable easier engagement.

If you determine that the flywheel ring gear needs to be replaced, see the engine manual for the appropriate procedure. If the starter pinion is worn, these can sometimes be replaced economically compared to the cost of replacing the whole starter. However, take into consideration the age and condition of the starter motor itself before attempting a pinion drive only replacement.

Task C5   **Inspect, clean, repair, and replace battery cables and connectors in the cranking circuit.**

Testing of starter circuit components using the voltage drop method was discussed in Task C1. This is the preferred method for determining the serviceability of cranking circuit components. Obvious physical damage such as cables with rubbed insulation and clamps with broken or cracked ends warrants replacement.

If the battery cable is OK, yet needs a new end(s), avoid using generic "bolt on" cable ends. They do not have the current carrying capability that a crimp or solder type terminal has, and will almost certainly cause trouble later with corrosion and high resistance. Use the appropriate size crimp or soldered terminal along with shrink tubing when repairing battery cables to minimize corrosion and resistance.

D. Charging System Diagnosis and Repair (8 Questions)

Task D1   **Diagnose dash-mounted charge meters and/or indicator lights that show a no charge, low charge, or overcharge condition; determine needed repairs.**

When an OEM indicator shows a charging system problem, it is important to understand how that instrument is wired into the system, and what to expect from the various types of warning devices.

Charge indicator lights generally illuminate with the key on-engine off, and should extinguish when the engine starts. Any other action usually indicates a problem with the charging system, but not always. Be sure you know how the system works and how the light is wired into the system before condemning an alternator.

Ammeters are gauges that indicate how much amperage is either going into or leaving the battery. These are no longer popular with manufacturers, as a small current loss from the battery in an undercharging situation would likely not be detectable by most ammeters.

Voltmeters are a more reliable indicator of charging system condition, along with the state of charge of the battery. If the voltmeter reads 13.5 to 14.5 volts, it is safe to say the alternator is charging. If it maintains that reading even under the heaviest of vehicle loads, you have effectively load tested the alternator at the same time. If battery voltage drops below 13.5 volts under any conditions with the engine running at high idle, the alternator is undercharging. Conversely, any reading exceeding specification means the system is overcharging. Charging voltage should not exceed 14.2 volts in electrical systems using gel cell batteries.

Task D2   Diagnose the cause of a no charge, low charge, or overcharge condition; determine needed repairs.

A no charge complaint could be caused by any number of different reasons, from a simple blown fuse to faulty brushes or diodes inside the alternator. Due to the many different types of systems, it is impossible to list all the potential causes. Get the service manual for the system you are working on and follow the troubleshooting procedures.

A low charge condition is one where the alternator charges properly with light loads, but falters under heavy ones. See Task D4 for details on diagnosing this complaint.

An overcharge condition (system operating over the maximum specified voltage) usually indicates a defective voltage regulator, but not always. In some systems, a defective "sense" diode in the alternator can send a low signal to the regulator, forcing it to overcompensate. Again, know the system you are working on before making assumptions.

Task D3   Inspect, adjust, and replace alternator drive belts/gears, pulleys, fans, and mounting brackets.

While worn drive belts are visually obvious, always determine the cause of the failure. Two of the more common causes are incorrect belt tension and misalignment. On vehicles without automatic tensioners, recheck belt tension after installing a new belt. Allowing the system to run for a while will seat the new belt, causing the need for readjustment. A belt that is loose will slip and fail prematurely.

Some alternators are shimmed fore and aft in their mounting brackets. Be sure to carefully align the alternator driven pulley with the engine drive pulley.

Task D4   Perform charging system voltage and amperage output tests; determine needed repairs.

Alternator output tests are simple procedures. All that is required is a battery load tester and a voltmeter. Before starting, look up the model ID on the alternator and find the specifications for maximum output. Also, check the drive belt for proper tension.

Attach the battery load tester across the battery as if you were going to load test it. However, for this test, attach the amp clamp (either the one with the load tester or a hand-held meter) around the alternator output wire. With the engine at high idle, load the battery tester down to a value slightly higher than the alternator rated output. This forces the alternator to output maximum amperage, which should be within 5 percent of specs. Note, however, that with the load tester drawing more current than the alternator can replace, system voltage will be down. This is normal. Next, reduce the current draw on the load tester to an amount slightly less than the alternator's rated output. System voltage should increase to between 13.5 to 14.5 volts.

Some systems can be tested for maximum output by a procedure known as full-fielding. This is a less desirable method than the former because if not careful, unregulated voltage can damage sensitive electrical/electronic components. If you are using this method, be sure to follow the manufacturer's instructions carefully.

Task D5 **Perform charging circuit voltage drop tests; determine needed repairs.**

For the alternator to provide maximum output, the output wire as well as the grounding of the alternator itself must be in good condition. This is checked by testing the voltage drop in each circuit.

Voltage drop in the alternator output wire can be tested in one of two ways. The first method involves performing the maximum output test (Task D4), and then measuring voltage drop between the output stud on the alternator and the positive battery cable. Perform this test using the same method as the starter circuit tests noted in Task C1. The reading should be less than one-tenth of a volt. Anything more indicates excessive resistance between those two points.

With the alternator at maximum output, test for voltage drop between the alternator housing and the battery ground terminal. This should also be one-tenth of a volt maximum.

A second method to determine voltage drop in the output circuit can be done with the engine not running. Connect the positive clamp of the load tester to the output stud on the alternator. Connect the negative clamp to the case. Draw current through the load tester equal to the alternator's maximum output. Do not exceed this voltage, or you may melt a wire. Test for voltage drop as described earlier.

Task D6 **Remove and replace alternator.**

Replacing an alternator is a simple job. Label wires before removing them. Be sure to remove the negative battery cable before removing the alternator wiring and after reinstalling the new alternator to prevent accidental sparks and possible wiring damage.

Task D7 **Inspect, repair, or replace connectors and wires in the charging circuit.**

In order for the alternator to transfer all of its power efficiently to the battery and the various loads in the vehicle, the wiring attached to the alternator must have clean connections and be in good condition. Other than a visual inspection for chafes, corrosion, etc., the best way to test the output wire is with the voltage drop test discussed in Task D5.

When repairing wiring in the charging system, be sure to use the proper size cabling and connectors for the circuit you are working on. Charts are available which list the recommended wire gauge size depending on amperage flow and length of wire. Also, use the proper crimp tools along with shrink tubing to ensure a solid and trouble free connection.

Task D8 **"Flash"/Full-field alternator to restore residual magnetism.**

"Flashing" the field is normally reserved for generators that lose their residual magnetism after sitting for long periods of time. This procedure is usually not necessary with alternators because many alternators rely on a key-switched signal to excite the circuit.

Full-fielding is a term used to describe a procedure where the field circuit of the alternator has full battery voltage applied to it. This is performed on systems with external voltage regulators where the regulator is suspect. Full-fielding bypasses the regulator and helps to isolate a problem. This procedure must be performed with care, however, because a full-fielded alternator will produce maximum output regardless of battery voltage. This can lead to battery boiling and component damage if the voltage is high enough.

E. Lighting Systems Diagnosis and Repair (6 Questions)

1. Headlights, Daytime Runnning Lights, Parking, Clearance, Tail, Cab, and Dash Lights (3 Questions)

Task E1.1 **Diagnose the cause of brighter than normal, intermittent, dim, or no headlight and daytime running light (DRL) operation.**

Multiple lights that are brighter than normal can only be caused by an alternator that is overcharging. Verify this with a check of alternator output using your DMM. A malfunctioning charging system can also cause dim light operation if the voltage is too low. While this would cause all the lights on the vehicle to be dim, verify this problem with your DMM also.

Dim lights can also be caused by problems such as excessive resistance in fuse holders, relays, wiring, switches, connectors, and chassis grounds. Of all the aforementioned possibilities, suspect the chassis grounds first. These cause the vast majority of problems related to this complaint, from simple loose hardware to poor metal-to-metal contact between chassis components. A simple voltage drop test (see Task A2) between the ground side of the bulb and the battery negative post will confirm this problem.

Intermittent operation can be caused by a cycling circuit breaker or a loose connection somewhere in the circuit. Often these circuit breakers are incorporated directly into the headlight switch. A cycling circuit breaker is caused by either a defective breaker or an overload in the system, and will usually produce a rhythmic "on-off" pattern. Loose connections are harder to find. A good tip would be to gently pull and wiggle suspect harnesses and connectors while watching the light action.

A blown fuse, defective circuit breaker, bad switch, or an open in the wiring generally causes no light operation. At this point, it is best to get a wiring diagram for the vehicle you are working on and probe with a test light at various points downstream from the power source until you have found the open.

Task E1.2 **Test, aim, and replace headlights.**

Headlight aim should always be checked on a level floor with the vehicle unloaded. In some states, the previous instructions may conflict with existing laws and regulations. If so, modify the instructions to meet the state's legal requirements.

To adjust headlights, first check headlight aim. Various types of headlight aiming equipment are available commercially. When using aiming equipment, follow the instructions provided by the equipment manufacturer.

When headlight aiming equipment is not available, aiming can be checked by projecting the upper beam of each light upon a screen or a chart at a distance of 25 feet ahead of the headlights.

Some manufacturers recommend coating the prongs and base of a new sealed beam with dielectric grease for corrosion protection. Use an electrical lubricant approved by the manufacturer.

Sealed-beam halogen headlights are designed to give substantially more light on high beam than incandescent, extending the driver's range of visibility for safer night driving. Halogen bulbs have the advantage of producing a whiter light, which helps improve visibility. Halogen bulbs also last longer, stay brighter, and use less wattage for the same amount of light produced. When replacing individual replaceable bulbs, avoid touching the glass envelope. Oil from the skin can cause the bulb to shatter when turned on.

Task E1.3 Test, repair, or replace headlight and dimmer switches, wires, connectors, terminals, sockets, relays, and control components.

Headlight switches are used to control the operation of the headlights and sometimes the parking and dash lights as well. Some manufacturers use separate toggle or rocker switches for each function, while others incorporate all switching functions in one multifunction switch. It is important to understand how a particular system is constructed to make troubleshooting easier. For this reason, always consult a wiring schematic for the vehicle you are working on.

Dimmer switches serve to switch current flow from either the low or high beam circuit. This switch can either be floor mounted or incorporated into the multifunction turn signal switch. The dimmer switch is a simple device that directs voltage to either the high or low beams. One wire is for power in, the other two direct current to either the high or low beam circuits.

Many headlight circuits contain relays to reduce the load conducted through the switch itself. A defective headlight or dimmer switch should affect both left and right headlights because they are wired in parallel. However, if only the high or low beam circuits are not operational, the headlight switch can usually be ruled out, because it is the dimmer switch that distributes power to the lights.

Task E1.4 Inspect, test, repair, or replace switches, bulbs, sockets, connectors, terminals, relays and wires of parking, clearance, and taillight circuits on trucks and trailers.

Parking, clearance, and taillight circuits can be controlled by a multifunction headlight switch, or they can be powered by separate toggle or rocker switches. When troubleshooting these lights, keep in mind that the most common causes of trouble are usually related to poor grounds, either at the lights themselves, or at a ground strap. Trailer connector plugs and sockets can be a source of problems when the pins corrode.

If only one clearance, taillight, or parking light is dim and the rest are OK, suspect a problem with that particular light or its ground. If all of the lights are dim, then assume that a ground or power supply malfunction is the cause.

Task E1.5 Inspect, test, repair, or replace dash light circuit switches, bulbs, sockets, connectors, terminals, wires, and printed circuits.

Dash illumination lights (not to be confused with the warning lights) on most trucks are fed from the same power source as the taillights and clearance lights on the vehicle, usually through the multifunction light switch assembly. However, between this source and the lights themselves will usually be a rheostat, also known as a dimmer switch, to allow the driver to reduce the intensity of the lights when driving at night. This dimmer switch may either be incorporated into the headlight switch, or be remote mounted elsewhere on the dash.

If the dash lights on a truck are dimmer than normal and all the other lights work OK, suspect a problem with the dimmer switch or its wiring, assuming that it has been adjusted to the correct position.

Task E1.6 Inspect, test, repair, or replace interior cab light circuit switches, bulbs, sockets, connectors, terminals, and wires.

When diagnosing problems related to interior cab lights, keep in mind that the door jam switches are on the ground side of the circuit. This enables the manufacturer to use a single wire to each switch. A short to ground on the switch side of the lamp will cause the lights to remain on constantly and not blow the fuse.

2. Stoplights, Turn Signals, Hazard Lights, and Backup Lights
(3 Questions)

Task E2.1   **Inspect, test, adjust, repair, or replace stoplight circuit switches, bulbs, sockets, connectors, terminals, relays, and wires.**

Stoplight circuits are relatively simple in that the major component is the switch itself. These are two types. On medium-duty trucks with hydraulic brakes, the switch is usually located on the brake pedal, activated by simple mechanical movement. On larger trucks with air brakes, the switch is incorporated into one of the brake application air lines. Air pressure acting against a diaphragm will close electrical contacts in the switch to complete the circuit. Some trucks will have two such switches: one for service brake applications, and another for parking brake applications.

Because stoplight and turn signal lamps share the same bulbs, the current from the stoplight switch is usually routed through the turn signal switch. This allows it to be directed properly when both the brakes and turn signals are activated at the same time.

Task E2.2   **Diagnose the cause of turn signal and hazard flasher light malfunctions; determine needed repairs.**

The turn signal circuit directs the output from the flasher unit to one side of the truck or the other depending on the position of the switch. If one side works properly but the other does not, assume that the flasher unit is OK and that the problem lies in the malfunctioning circuit. If the problem exists in the trailer and not the truck, a good place to start troubleshooting would be at the trailer connector plug. At this point each circuit can be split up and diagnosed.

The brake lamp switch can also be considered part of the turn signal circuit, because if the brakes are applied during a turn, the side that the truck is being turned to must still be able to flash. This is accomplished by routing the current from the stoplight switch directly into the turn signal switch. At this point, the turn signal switch will direct the current flow in the proper direction depending on position.

Task E2.3   **Inspect, test, repair, or replace turn signal and hazard circuit flashers, switches, bulbs, sockets, connectors, terminals, relays, and wires.**

Turn signal flashers that operate faster or slower than normal can indicate higher or lower than normal current demands on the circuit, respectively. For example, if a bulb should burn out on one side of a vehicle, this would reduce current demand in that particular circuit, causing the flasher assembly to blink slower than normal. Conversely, if a condition caused higher than normal current flow in the circuit, say due to extra lights being added, current flow would be increased, causing the bi-metal contacts inside the flasher to heat up and cycle at a faster rate. This condition could cause premature failure of the flasher assembly.

Newer vehicles with electronic flashers will incorporate a relay in the circuit to handle the high-current switching demands. This type of system is not as affected by either defective or additional lights in the circuit.

Task E2.4   **Inspect, test, adjust, repair, or replace backup light and warning device circuit switches, bulbs, sockets, horns, buzzers, connectors, terminals, and wires.**

Switches mounted on the transmission or shift linkage usually control backup lights. When the transmission is shifted into reverse, contacts inside the switch are closed and the backup circuit is energized.

Warning alarms are installed on some trucks, especially inner-city delivery vehicles, to alert persons near the truck that it is about to back up. These devices will use the same switch and signal that the backup lights do, but will also incorporate a relay in the circuit to handle the high power demands of the alarm to avoid overloading the backup circuit.

F. Gauges and Warning Devices Diagnosis and Repair (6 Questions)

Task F1   **Diagnose the cause of intermittent, high, low, or no gauge readings; determine needed repairs. (Does not include charge indicators.)**

When troubleshooting gauge systems, first determine whether the problem exists with all the instrumentation or just isolated gauges. For example, older style bi-metal gauges are sensitive to voltage fluctuations. If you find that they all read too high or too low, it is possible that the voltage regulator is malfunctioning. The purpose of the voltage regulator is to maintain a steady voltage supply to the gauges, regardless of battery voltage. Individual gauges that read erratically may be caused by any number of factors, including high resistance, a defective sending unit, or a malfunctioning gauge. Follow manufacturer's instructions for out-of-dash gauge testing.

Later model instrumentation systems primarily use magnetic style gauges. Most are not affected by changes in system voltage, and therefore do use a separate instrument voltage regulator.

Not all gauges in an instrument panel are electrically operated. Some gauges, such as air application gauges, are mechanically actuated and not affected by electrical problems. Air application gauges must test within 4 psi of an accurate master gauge.

Never rely only on the accuracy of dash gauges to make major decisions such as an engine rebuild. For example, if vehicle instrumentation indicates low engine oil pressure, confirm the problem with a known accurate, mechanical gauge to verify the condition.

Task F2   **Diagnose the cause of data bus driven gauge malfunctions; determine needed repairs.**

Instrumentation systems in current vehicles rely on information sourced from sensors that signal information to an engine or chassis control module. From there, the module will broadcast that information over a data bus to the other on-board computers. One of these computers can be dedicated specifically to the instrumentation. It is the function of the instrumentation module to take these digital signals from the main ECM and display them on the dash in an analog, bar graph, or digital format, depending on vehicle options.

Some truck gauges, such as transmission or rear axle temperature gauges, rely on information sourced from a module other than the engine computer, since the engine ECM does not need this type of information. Keep this in mind when troubleshooting such systems.

Newer instrumentation systems will go into a self-test mode when the key is first turned on to verify the operation of each gauge. When the accuracy of any gauge is suspect, first determine if the dash is receiving accurate information. To do this, connect the appropriate service tool to the vehicle diagnostic connector and scroll the menu options to scan the various sensor values that the module is receiving. From there you can determine where a problem lies by checking the values against actual vehicle operating parameters.

Task F3   **Inspect, test, adjust, repair, or replace gauge circuit sending units, gauges, connectors, terminals, and wires.**

To test sending units, use the service manual for the vehicle to determine what the senders should read at certain operating parameters. For example, a certain resistance value might be specified for a fuel level sending unit when the tank is half full. Or, a temperature sending unit could be tested in boiling water, again, with a specified resistance value at that point. Another quick way to check thermister type temperature senders on an electronic vehicle is to compare values (using a diagnostic tool) to those specified in the OEM software parameters.

Many electrical gauges and some electronic ones can be checked using a variable resistance test box. Unplug the wire at the sending unit and substitute the resistance of the test box while comparing the action of the gauge against manufacturer specifications.

Some gauges such as pyrometers do not use a separate power source. Since their purpose is to measure high temperatures such as exhaust gases, a signal is derived from the sender probe using heat energy to generate a small voltage which is correlated with a temperature value at the dash display.

Task F4   Inspect, test, repair, or replace warning devices (lights and audible) circuit sending units, bulbs, audible component, sockets, connectors, terminals, wires, and printed circuits/control modules.

Warning lights and buzzers on older vehicles (pre-electronic) have a sender or switch on the ground side of a particular monitored function that will close and allow the circuit to be powered up in the event of a malfunction.

Many newer vehicles have built-in warning systems designed to alert the driver in case of a malfunction that might otherwise damage the engine or other components. Usually, an audible alarm unit is located behind the dash and will sound when a malfunction is detected. Often they are incorporated into the dash itself and not serviced separately; however, some are remote mounted and can be replaced.

Most trucks with electronically controlled engines will also have at least two dash warning lights to indicate faults. One will be a "check engine" light, and the other a "stop engine" light. The former will alert the driver to a condition that needs attention at the driver's earliest convenience, while the latter will indicate a problem that requires immediate engine shutdown, such as low oil pressure or high water temperature. Depending on the configuration of the engine protection system (usually programmable), the stop engine light may also be accompanied by either a ramp down in engine power and/or an automatic engine shutdown.

When testing circuits in an electronic instrumentation system, always use a DMM to check voltage and never test lights. Test lights draw too much current from an electronic circuit to make this type of testing valid and many damage the circuit. It should be noted that most electronic circuits operate at voltages lower than battery voltage.

Task F5   Inspect, test, replace and calibrate electronic speedometer, odometer, and tachometer systems.

Most speedometers, odometers, and tachometers on current trucks are driven electronically through information sent via the data bus (see Task F2). When the information on any of these indicators is suspect, first check the calibration menu of the manufacturer diagnostic software tool to see how the vehicle speed parameters are set up. For example, in order for the computer to accurately display speed or mileage figures, it must first know the rear axle ratio and the tire size. Some engine management systems even track tire wear to adjust the speedometer reading. If these parameters are not properly entered into the computer, inaccurate data will result. These data fields may be password protected.

Many speedometers and tachometers get their information through sensors known as magnetic pickups. These are simple devices that set up a magnetic field near a toothed wheel. As the wheel rotates and the teeth are driven through the magnetic field, a small voltage signal is generated in the pickup. Signal frequency is used by the ECM to determine engine or wheel speed.

Some older vehicles with electric gauges can derive the tachometer signal off the alternator, usually the "R" terminal. If these gauges are suspect, an easy way to test them is with the use of a signal generator. This device simulates the signal that would normally be fed off a magnetic pickup or alternator phase tap. Using the manufacturer repair manual, the frequency of the signal being sent to the gauge can be correlated with the actual reading.

G. Related Systems (5 Questions)

Task G1   **Diagnose the cause of constant, intermittent, or no horn operation.**

Because an electric horn uses high current, the circuit will almost always include a relay. If the horn does not work at all, first check for power going to the horn itself. If the unit is properly grounded and does not operate, the unit is defective and should be replaced. Check both the control and power circuits at the relay.

Task G2   **Inspect, test, repair, or replace horn circuit relays, horns, switches, connectors, terminals, and wires.**

When testing a horn relay circuit, first check for fused power going to the positive side of the horn relay (usually pin #30). Next, check for a signal from the steering wheel switch. This may be either positive or ground side switching, depending on the manufacturer. Assuming the use of a 5 pin mini-relay, the circuit should function exactly as described in Task A9.

Task G3   **Diagnose the cause of constant, intermittent, or no wiper operation; diagnose the cause of wiper speed control and/or park problems.**

Most electric windshield wiper systems use a 2-speed motor. This motor will either be a permanent magnet type with high and low speed brushes, or it will have an external resistor pack much like heater blower motors. Inside the first style of motor will be a low speed and a high speed brush set. If the motor works OK in one speed but is sluggish or not functioning in the other, it may be due to defective brushes inside the motor. First ensure that the proper voltage is coming from the appropriate side of the wiper switch with the circuit closed. Consult the appropriate manufacturer's repair manual for the wiring diagram of the vehicle.

If a motor fails to work at all with power applied, remember that some motors incorporate thermal overload protection should a motor overheat due to binding linkage (see Task G5).

Task G4   **Inspect, test, repair, or replace wiper motor, resistors, park switch, relays, switches, connectors, terminals, and wires.**

If wipers should fail to park properly after the switch is turned off, it may be due to a defective park switch usually incorporated into the motor itself. The function of this switch is to continue motor operation until the blades are fully retracted. Also, check that the park switch is receiving power when the wiper switch is in the "off" position.

Task G5   **Inspect and replace wiper motor transmission linkage, arms, and blades.**

If a wiper system parks consistently in the wrong position, it may be due to misadjusted wiper crank arm linkage. Readjust, using factory supplied indexing marks where applicable.

If wiper action is sluggish or intermittent, it may be caused by binding linkage. Disconnect the motor crank arm and manually move the wiper linkage to be sure that the mechanisms are free to operate without undue force.

Task G6   **Inspect, test, repair, or replace windshield washer motor or pump/relay assembly, switches, connectors, terminals, and wires.**

Most washer pump circuits are relatively simple ones. The switch for the pump is almost always incorporated into the wiper control switch, and will control the washer pump through either positive or ground side switching. If the wipers work but the washers do not, you can eliminate a fuse or circuit breaker problem since the wiper switch assembly usually draws power from the same source.

Task G7   **Inspect, test, repair, or replace sideview mirror motors, heater circuit grids, relays, switches, connectors, terminals, and wires.**

Moveable mirror assemblies incorporate two different motors, one for horizontal movement and the other for vertical. If one or both motors do not operate, first confirm that power exists at the motor terminals when the appropriate side of the mirror switch is depressed.

Heated mirrors can be considered a safety option in cold climates. When testing a heater circuit for proper operation, be aware that some circuits contain a fast warmup cycle, which then drops considerably in current draw as the mirror warms.

Task G8   **Inspect, test, repair, or replace heater and A/C electrical components including: A/C clutches, motors, resistors, relays, switches, controls, connectors, terminals, and wires.**

Heater blower motors are typically 3 or 4 speed. To accomplish this, the current flow from the switch is routed through a set of resistors wired in series. If one of these resistors should fail, the motor may work in high speed only, since the resistors are bypassed in the high-speed position. Blower motor resistors are usually found in the blower air stream to keep them from burning up. Even though this voltage drop is intentional, it still produces heat.

AC compressor clutches can be controlled by either a set of manual switches and pressure sensors, or directly from an AC system microprocessor. Research how the system is configured before troubleshooting.

When replacing AC compressor clutches, many need to be shimmed properly to maintain proper air clearance in the disengaged position. This must be done carefully. Too little clearance may cause the clutch to drag and burn up; too much might prevent engagement.

Task G9   **Inspect, test, repair, or replace cigarette lighter and/or auxiliary power outlet case, integral fuse, connectors, terminals, and wires.**

Many vehicles today have what are known as auxiliary power outlets in place of or in addition to cigarette lighters. These are designed as convenience items for drivers should they need to plug in portable accessories such as a laptop computer.

Some of these outlets look like cigarette lighter sockets, while others are multiple pin connectors designed to provide both key switch operated (on-off) and constant power. When troubleshooting these devices, remember that these are at least two different circuits.

Task G10   **Diagnose the cause of slow, intermittent, or no power window operation.**

If a power window operates sluggishly or not at all, it may be due to either incorrect voltage being supplied to the motor, or a binding window mechanism. To differentiate between the two, check for the proper voltage at the motor when the switch is actuated. If the voltage is correct and/or the system continually blows fuses, binding linkage is most likely the cause. A stalled or nearly stalled electric motor will almost always cause a current overload.

Task G11   **Inspect, test, repair, and replace motors, switches, relays, connectors, terminals, and wires of power window circuits.**

If you find that both windows in a truck are inoperative, a good place to start troubleshooting is to check the power source. If one side works but the other does not, check the switch, followed by testing for proper voltage at the motor.

When testing a power window motor circuit, keep in mind that the motors are reversible. This means that the wiring connected to the motor will not have a dedicated power and ground assignment. Rather, this will be alternated at the control switch depending on which direction the switch is activated.

Task G12   **Inspect, test, repair, and replace cruise control electrical components.**

Many current trucks use cruise control systems managed by the engine computer. Since the ECM controls the speed of the engine by varying the fueling rate, cruise control is a natural extension of its capabilities. The ECM can also engage the engine compression brake automatically on downhill runs while in cruise control.

If the cruise control does not operate properly, check that the control switches are all working as they should. This can be tested easily with most troubleshooting software by monitoring switch operation (open or closed) via the diagnostic tool. If this fails to resolve the problem, check the vehicle calibration display to be sure that the operating parameters were set up correctly.

Task G13   **Inspect, test, repair, or replace engine cooling fan electrical control components.**

Most fans on over-the-road trucks are air-operated devices that obtain their air signal from a 12-volt solenoid. This solenoid is actuated by some sort of temperature sensing switch on older vehicles, or is controlled by the engine computer on newer vehicles. Find out which type you are working on before troubleshooting.

On systems controlled by a remote temperature switch, this switch will either control the power or ground side of the circuit. When it makes the internal connection, the air solenoid is energized, and air is directed or exhausted to either engage or disengage the fan, depending on the type of fan and solenoid design.

On ECM controlled fan circuits, it is important to understand the control system logic for the engine/vehicle you're working on. In addition to the normal on-off operation described earlier, some control systems are programmed to engage the fan whenever a temp sensor on the engine fails, regardless of its reading. This is designed to be a fail-safe protection feature for the engine, since the module cannot determine the actual temperature. Other systems engage the fan whenever any portion of the circuit fails, by a method known as reverse ground side switching. Again, this is a fail-safe feature meant to protect the engine in the event of a problem with the electrical system. Remember that trucks may engage the fan during cruise control set speed overrun (downhill driving) to assist in vehicle retardation.

Sample Test for Practice

Sample Test

Please note the letter and number in parentheses following each sample question. They match the overview in section 4 that discusses the relevant subject matter. You may want to refer to the overview using this cross-referencing key to help with questions posing problems for you.

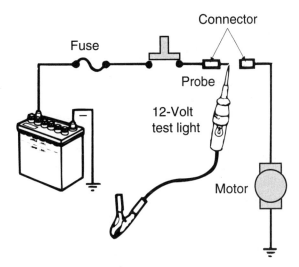

1. In the figure shown above, the battery is fully charged and the motor is inoperative. When using a test lamp to check the circuit for voltage, the technician should connect the ground clip to which of these areas?
   A. The fuse
   B. The battery positive terminal
   C. The battery negative terminal
   D. The battery (hot) side of the switch (A2)

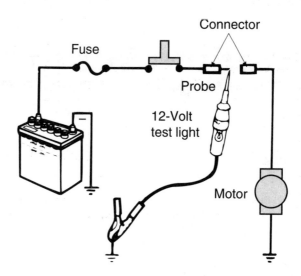

2. In the figure shown above, technician A says the circuit is open. Technician B says to diagnose this circuit using an ohmmeter. Who is right?
   A. Technician A only
   B. Technician B only
   C. Both A and B
   D. Neither A nor B   (A1)

3. Which of the following is NOT used when checking a circuit for continuity?
   A. An ohmmeter
   B. An ammeter
   C. A voltmeter
   D. A test lamp   (A1)

4. After removing a 30-amp circuit breaker from a fuse panel a technician checks continuity across its terminals with a digital multimeter (DMM). Technician A says that the current from the ohmmeter will open the circuit breaker. Technician B says the ohmmeter should read infinity when the circuit breaker is good.
   Who is right?
   A. Technician A only
   B. Technician B only
   C. Both A and B
   D. Neither A nor B   (A4)

5. While diagnosing a truck with electronic engine management and a "no-start" complaint, technician A says you should only use a digital multimeter to check voltages on a control module circuit. Technician B says that the digital multimeter should have a 10 megohm or higher impedance. Who is right?
   A. Technician A only
   B. Technician B only
   C. Both A and B
   D. Neither A nor B   (A2)

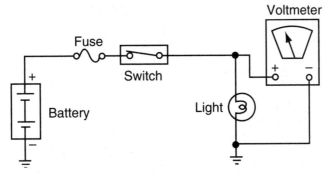

6. In the figure shown above, the battery is fully charged and the switch is closed. With the voltmeter connected as shown, a reading of 0 volts indicates:
   A. the fuse is open.
   B. the bulb is bad.
   C. the bulb is OK.
   D. the voltmeter leads are reversed. (A2)

7. When using a voltmeter to perform a voltage drop test in a circuit, the leads should be connected in what way?
   A. To the battery terminals
   B. From the positive battery terminal to ground
   C. In series with the circuit being tested
   D. In parallel with the circuit being tested (A2)

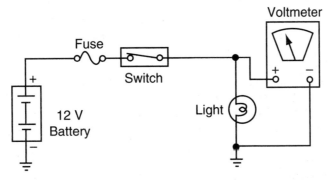

8. The battery in the figure above is fully charged and the switch is closed. The voltage drop across the light bulb indicated on the voltmeter is 9 volts. Technician A says there may be a high resistance problem in the light bulb. Technician B says the circuit may be grounded between the switch and the light. Who is right?
   A. Technician A only
   B. Technician B only
   C. Both A and B
   D. Neither A nor B (A2)

9. Technician A says an ammeter should be used to check for a short circuit between circuits. Technician B says to make sure you fully charge the battery before checking a circuit for current draw. Who is right?
   A. Technician A only
   B. Technician B only
   C. Both A and B
   D. Neither A nor B (A3)

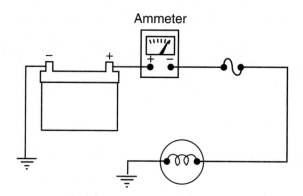

Ammeter

10. In the figure above, the ammeter indicates current flow through the bulb is higher than specified. The cause of this high current could be which of the following?
    A. The fuse has an open circuit.
    B. The battery voltage is low.
    C. The light bulb filament is shorted.
    D. The bulb filament has high resistance.   (A3)

11. Technician A says an ammeter is used to test continuity. Technician B says an ammeter indicates current flow in a circuit. Who is right?
    A. Technician A only
    B. Technician B only
    C. Both A and B
    D. Neither A nor B   (A3)

12. Two technicians are discussing resistance measurement with an ohmmeter. Technician A says you can connect an ohmmeter in a circuit in which current is flowing. Technician B says when testing a spark plug wire with 20,000 ohms resistance, use the ×100 meter scale. Who is right?
    A. Technician A only
    B. Technician B only
    C. Both A and B
    D. Neither A nor B   (A4)

13. The blower motor in a truck is running very slowly. An ammeter shows a low-current flow. This could be caused by:
    A. high resistance in the circuit.
    B. low resistance in the circuit.
    C. an overcharged battery.
    D. a shorted blower motor.   (A3)

14. When testing a circuit for voltage drop, what has to occur to obtain an accurate reading?
    A. There must be voltage present somewhere in the circuit being tested.
    B. There must be resistance in the circuit being tested.
    C. There must be current flow in the circuit being tested.
    D. The tested leads must be probed in series with the circuit being tested. (A2)

15. Technician A says that unwanted resistance in an electrical circuit will always produce heat. Technician B states that unwanted resistance will cause reduced current flow. Who is right?
    A. Technician A only
    B. Technician B only
    C. Both A and B
    D. Neither A nor B   (A2, C1)

16. Technician A says that a DMM can be used to test current flow directly through the meter in any electrical circuit. Technician B states that to test for high amperage flow, it is best to use a current clamp to prevent damage to the meter. Who is right?
    A. Technician A only
    B. Technician B only
    C. Both A and B
    D. Neither A nor B   (A3)

17. When using a 12V test lamp to test for voltage, all of the following are true **EXCEPT:**
    A. the test lamp may be connected to the battery ground.
    B. the battery should first be disconnected.
    C. the test lamp may be connected to a chassis ground.
    D. a weatherproof connector may be backprobed.   (A2)

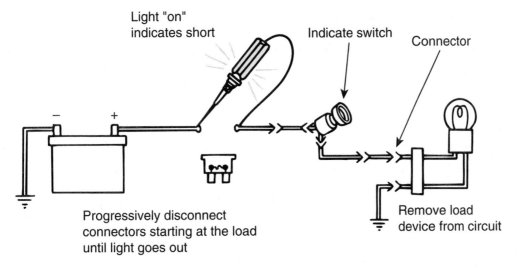

Light "on" indicates short

Indicate switch

Connector

− +

Progressively disconnect connectors starting at the load until light goes out

Remove load device from circuit

18. The light bulb in the figure above is inoperative. A 12-volt test light is installed in place of the fuse, the test light is on, and the bulb is off. When the connector near the bulb is disconnected, the 12-volt test light remains on. Technician A says the circuit may be shorted to ground between the fuse and the disconnected connector. Technician B says the circuit may be open between the disconnected connector and the bulb. Who is right?
    A. Technician A only
    B. Technician B only
    C. Both A and B
    D. Neither A nor B   (A5)

19. Technician A says that a self-powered test lamp can be used to check for continuity in a circuit managed by an electronic control module. Technician B states that an analog multimeter may be used to test voltages in an electronic circuit. Who is right?
    A. Technician A only
    B. Technician B only
    C. Both A and B
    D. Neither A nor B   (A1)

20. A customer has a truck towed to the shop and says that the starter would not crank the engine. What should be checked first?
    A. Ground cable connection
    B. The starter solenoid circuit
    C. The ignition switch crank circuit
    D. The battery for proper charge   (B2)

21. While performing a battery drain test, the ignition switch should be in the:
    A. accessory position.
    B. run position.
    C. crank position.
    D. off position.   (A6)

22. Technician A says that a current draw of 2 amps could cause the battery to discharge. Technician B says that two amps is a normal key-off draw for a vehicle with an electronic control module. Who is right?
    A. Technician A only
    B. Technician B only
    C. Both A and B
    D. Neither A nor B   (A6)

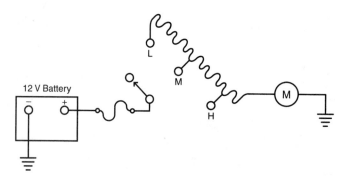

23. The blower motor in the circuit shown above draws 12 amps in the high speed position. How many amps would you expect it to draw in the medium speed position?
    A. Less than 12
    B. More than 12
    C. 12
    D. None   (A3)

24. While checking the fuses in a tractor/trailer, the technician finds an open fuse. The next step is to:
    A. replace the fuse with the next higher amperage rating.
    B. check the affected circuit for a short to ground.
    C. check the affected circuit for an open.
    D. install a circuit breaker of the same amperage rating as the fuse.   (A7)

25. When replacing a fusible link, technician A says you should disconnect the battery first. Technician B says you should always use the same gauge fuse link wire as the circuit being repaired. Who is right?
    A. Technician A only
    B. Technician B only
    C. Both A and B
    D. Neither A nor B   (A7)

26. Technician A says that any kind of wire can be used to replace a faulty fusible link as long as the gauge is one size smaller than the circuit being protected. Technician B says that a circuit breaker should be replaced after repairing a short in a circuit. Who is right?
    A. Technician A only
    B. Technician B only
    C. Both A and B
    D. Neither A nor B   (A7)

27. Maxi-fuses are:
  A. fuses of a higher quality than standard ones.
  B. fuses that are bigger than standard and used in place of fusible links.
  C. another name for a circuit breaker.
  D. usually found in all truck electrical circuits because of their large current
      requirements.   (A7)

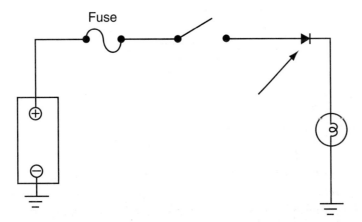

28. The arrow in the figure shown above is pointing to a:
  A. diode.
  B. resistor.
  C. capacitor.
  D. thermistor.   (A8)

29. Technician A says when the switch is closed in the figure shown above, the light
    bulb will illuminate normally. Technician B says when the switch is closed the
    circuit will short, causing the fuse to burn out. Who is right?
  A. Technician A only
  B. Technician B only
  C. Both A and B
  D. Neither A nor B   (A8)

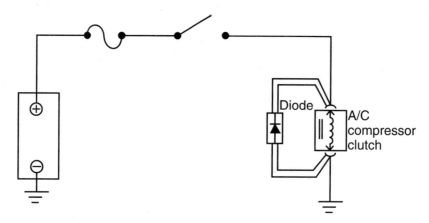

30. The diode in the figure shown above is being used to:
    A. allow the A/C compressor to run in one direction only.
    B. protect the A/C compressor clutch from a voltage spike.
    C. protect the circuit from a possible voltage spike.
    D. limit the current flow to the A/C compressor clutch.   (A8)

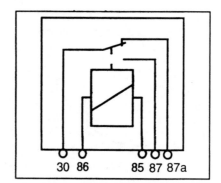

31. A 5 pin mini-relay shown in the figure above has each pin identified with a number. What is pin #30 for?
    A. Control power in
    B. Control ground
    C. High amperage (supply) power in
    D. High amperage (supply) power out, normally closed   (A9)

32. Technician A says that relays are never used by electronic control modules to control engine components. Technician B says that solenoids are found only on starter motors. Who is right?
    A. Technician A only
    B. Technician B only
    C. Both A and B
    D. Neither A nor B   (A9)

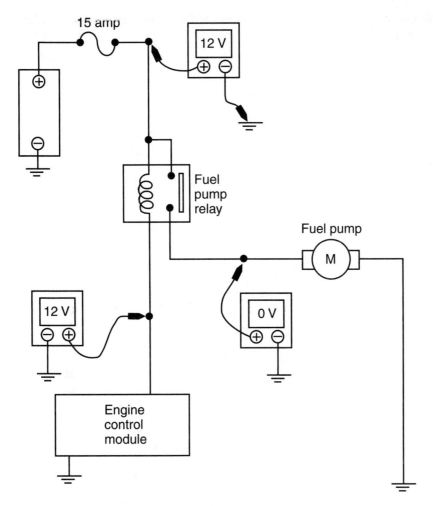

33. In the fuel pump circuit shown in the figure above, the fuel pump is inoperative. The most likely cause is:
    A. an open 15 amp fuse.
    B. a defective engine control module.
    C. a defective fuel pump relay.
    D. a defective fuel pump.   (A9)

34. When checking open circuit battery voltage, technician A says that a 12-volt battery is considered fully charged if a voltmeter probed across it reads anything over 12 volts. Technician B states that the battery must read at least 13.5 to 14.5 volts to be considered fully charged. Who is right?
    A. Technician A only
    B. Technician B only
    C. Both A and B
    D. Neither A nor B   (B2)

35. At 80°F, what is the correct specific gravity of electrolyte in a fully charged battery?
    A. 1.200 to 1.220
    B. 1.220 to 1.260
    C. 1.260 to 1.280
    D. 1.280 to 1.300   (B2)

36. A maintenance-free battery is low on electrolyte. Technician A says a faulty voltage regulator may cause this problem. Technician B says a loose alternator belt may cause this problem. Who is right?
    A. Technician A only
    B. Technician B only
    C. Both A and B
    D. Neither A nor B   (D2)

37. When discussing a battery capacity test with the battery temperature at 70°F, Technician A says the battery discharge rate is calculated by multiplying two times the battery reserve capacity rating. Technician B says the battery is satisfactory if the voltage remains above 9.6 volts under load at 70°F. Who is right?
    A. Technician A only
    B. Technician B only
    C. Both A and B
    D. Neither A nor B   (B1)

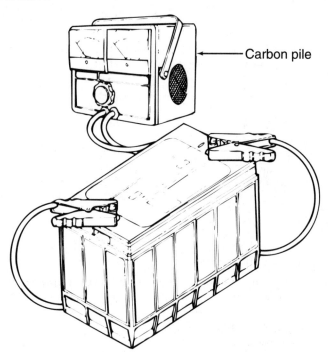

— Carbon pile

38. In the figure shown above, what test is being performed?
    A. A battery drain test
    B. A battery capacity test
    C. A battery voltage test
    D. A state of charge test   (B1)

39. Technician A says when performing a battery load test on a 12-volt battery, a good battery should have a voltage reading below 9.6 volts while under load. Technician B says the battery should be discharged at 2 times its ampere-hour rating. Who is right?
    A. Technician A only
    B. Technician B only
    C. Both A and B
    D. Neither A nor B   (B1)

40. While performing an open circuit voltage test, a reading of 12.4 volts is obtained. Technician A says this indicates a low battery. Technician B says the battery should be replaced. Who is right?
    A. Technician A only
    B. Technician B only
    C. Both A and B
    D. Neither A nor B   (B3)

41. Two technicians are discussing key-off battery drain problems. Technician A states that no amount of key-off battery drain is acceptable on any truck. Technician B says battery drain can be caused by excess moisture on top of the battery. Who is correct?
    A. Technician A only
    B. Technician B only
    C. Both A and B
    D. Neither A nor B   (A6, B3)

42. All of the following are acceptable battery and cable maintenance procedures **EXCEPT:**
    A. remove the negative battery cable last and reinstall first to avoid sparks.
    B. clean corrosion and moisture accumulation on the battery top with water and baking soda solution.
    C. only replace battery cable ends with proper solder or crimp on terminals and heat shrink tubing.
    D. coat battery terminal ends with a protective grease to retard corrosion.   (B3)

43. Technician A says that battery holddowns should always be installed to prevent batteries from excessive bouncing and possible internal damage. Technician B states that a battery box need not be cleaned when replacing a battery because the case is insulated and the battery cannot discharge because of it. Who is right?
    A. Technician A only
    B. Technician B only
    C. Both A and B
    D. Neither A nor B   (B4)

44. A technician is attempting to charge a battery and yet it apparently will not accept a charge according to the ammeter on the charger. What is LEAST likely to be the problem?
    A. The battery is already fully charged.
    B. The battery is highly sulfated.
    C. Poor contact exists between the charging clamp and the battery post.
    D. Excessive moisture accumulation across the top of the battery is causing surface discharge.   (B5)

45. Before replacing a battery, it is important to:
    A. replace the battery cables and terminals.
    B. check the charging system.
    C. replace the starter motor.
    D. replace the alternator belt.   (B1)

46. Technician A says that a low battery cannot generate explosive vapors to the extent that a fully charged battery can, and therefore an explosion is far less likely. Technician B states that it is good practice to wear eye protection when jump-starting. Who is right?
    A. Technician A only
    B. Technician B only
    C. Both A and B
    D. Neither A nor B   (B6)

47. When cleaning or servicing the battery, cables, holddown, and tray, you should:
    A. use an air nozzle to blow off the components.
    B. use a cleaning solvent such as mineral spirits.
    C. inspect and clean the terminals if needed.
    D. use sulfuric acid to dissolve the residues.   (B3)

48. When testing a battery for open circuit voltage, technician A says that this should be done immediately after the battery comes off of the charger. Technician B states that this can only be done immediately after the batteries have been charged in the vehicle. Who is right?
    A. Technician A only
    B. Technician B only
    C. Both A and B
    D. Neither A nor B   (B2)

49. When charging a battery, you should never:
    A. disconnect the negative battery cable first.
    B. charge the battery until it reads 1.265 specific gravity.
    C. reduce the fast-charging rate when specific gravity reaches 1.225.
    D. charge a frozen battery.   (B5)

50. Which of the following is the most acceptable way to recharge a battery that is very low?
    A. Fast charge rate
    B. Slow charge rate
    C. Using the vehicle's charging system
    D. Adding additional electrolyte   (B5)

51. While jump-starting a vehicle with a booster battery, technician A says the accessories should be turned on in the booster vehicle while starting the vehicle being boosted. Technician B says the negative booster cable should be connected to an engine ground on the vehicle being boosted. Who is right?
    A. Technician A only
    B. Technician B only
    C. Both A and B
    D. Neither A nor B   (B6)

52. A medium-duty truck with a dead battery is being jump-started. Technician A says the engine should be running on the boost vehicle before trying to start the dead vehicle. Technician B says the engine should be off while connecting the booster cables. Who is right?
    A. Technician A only
    B. Technician B only
    C. Both A and B
    D. Neither A nor B   (B6)

53. Which of the following is found in the starting circuit component?
    A. A solenoid
    B. A ballast resistor
    C. A voltage regulator
    D. An electronic control module (ECM)   (C3)

54. A starter circuit voltage drop test checks everything **EXCEPT:**
    A. battery voltage.
    B. resistance in the positive battery cable.
    C. resistance in the negative battery cable.
    D. condition of the solenoid internal contacts.   (C1)

55. A starter solenoid makes a loud clicking noise when the key switch is turned to the start position, but the starter motor fails to rotate. A check of battery voltage finds the battery is fully charged. Which of the following is most likely the problem?
    A. Defective magnetic switch
    B. Defective key switch
    C. Defective internal solenoid contacts
    D. An open circuit between the magnetic switch and the starter solenoid   (C3)

56. How is the starter ground circuit resistance check performed?
    — A. A voltmeter is connected between the ground terminal of the battery and the base of the starter and read while the engine is being cranked.
    B. An ohmmeter is connected between the starter relay housing and the starter housing.
    C. An ohmmeter is connected between the ground side of the battery and the starter housing and read while the engine is cranking.
    D. A voltmeter is connected between the positive side of the battery and the starter solenoid while the engine is off.   (C2)

57. An alternator is overcharging. Technician A says that this can only be caused by a defective voltage regulator. Technician B states that this can be caused by excessive resistance in the charging circuit wiring. Who is right?
    A. Technician A only
    B. Technician B only
    C. Both A and B
    D. Neither A nor B   (D2)

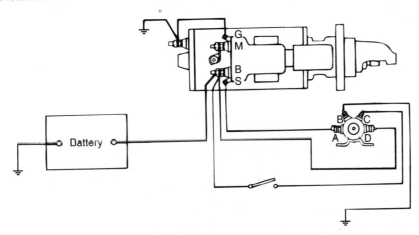

58. A technician is to perform a voltage drop test across the starter solenoid internal contacts. Using the figure shown above, where should he probe his voltmeter leads?
    A. Between the positive battery terminal and point B
    B. Between the positive battery terminal and point M
    C. Between points B and M
    D. Between points G and ground   (C3)

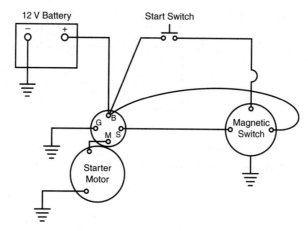

59. Using the figure shown above, when the starter circuit is activated the magnetic switch clicks, but the starter does not operate. The technician removes the wire at point S and checks for voltage with the key switch activated. The magnetic switch clicks and the technician finds 12 volts at the wire. The starter and solenoid assembly have been bench tested and found to be in working order. What could be the problem?
    A. The magnetic switch has a poor ground.
    B. The contacts inside the magnetic switch are badly corroded.
    C. There are faulty contacts inside the key switch.
    D. The negative battery cable has a poor ground connection at the block.   (C2)

60. In the figure above question 59, technician A says that the magnetic switch is responsible for transmitting all of the current flow to the starter motor. Technician B states that the key switch transmits all of the current to the starter solenoid. Who is correct?
    A. Technician A only
    B. Technician B only
    C. Both A and B
    D. Neither A nor B   (C2)

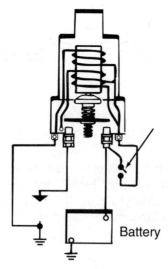

61. In the figure shown above, what does the arrow point to?
    A. The start switch
    B. The relay switch
    C. Open pull-in winding
    D. Open hold-in winding   (C2)

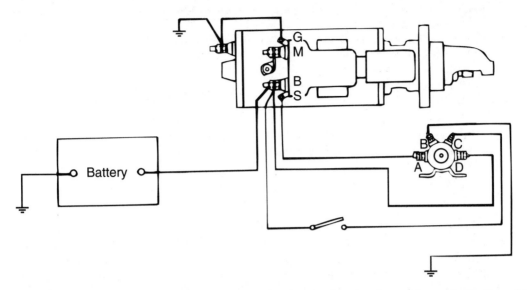

62. What component is being checked in the figure shown above if terminals C and D are jumped?
    A. The starting switch
    B. The battery
    C. The starter solenoid
    D. The starter   (C2)

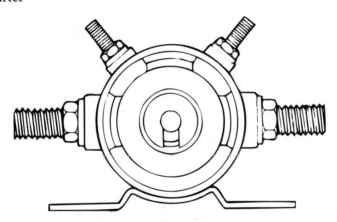

63. What is pictured in the figure shown above?
    A. A magnetic switch
    B. A starting-safety switch
    C. A starting switch
    D. A starter solenoid   (C2)

64. A truck with a starting-safety switch is brought in for service. Technician A says that the starting-safety switch is used to prevent vehicles with automatic transmissions from being started in gear. Technician B states that starting-safety switches are often called neutral safety switches. Who is right?
    A. Technician A only
    B. Technician B only
    C. Both A and B
    D. Neither A nor B   (C2)

65. All of these are part of the control circuit **EXCEPT:**
    A. a starting switch.
    B. the battery.
    C. a starting-safety switch.
    D. a magnetic switch.   (C2)

66. A starter motor spins but fails to turn the engine over. Technician A says that the starter solenoid may not be activated. Technician B says that the starter drive may be faulty. Who is right?
    A. Technician A only
    B. Technician B only
    C. Both A and B
    D. Neither A nor B   (C3)

67. All of these conditions would cause the starter not to start the engine **EXCEPT:**
    A. the battery does not connect battery power to the starter motor.
    B. the solenoid does not pull the starter drive pinion in mesh with the engine flywheel.
    C. failure of the low-current circuit to switch large-current circuit.
    D. the starter drive pinion fails to retract from the flywheel.   (C3)

68. When checking cranking circuit voltage loss on the insulated side of the starting circuit, the technician uses a digital multimeter on the voltage setting and performs which of these?
    A. Connect one test lead to the positive side of the battery cable while the starter is turning
    B. Connect one test lead to the starter body while the starter is not running
    C. Connect one lead to the negative side of the battery cable while the starter is turning
    D. Allow the vehicle to warm up before beginning checks   (C1)

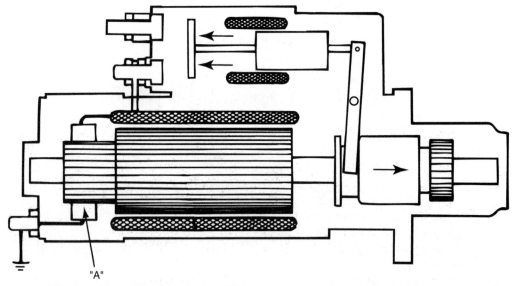

"A"

69. The item indicated by the arrow labeled "A" in the figure shown above, if defective, will cause which of these conditions?
    A. The cranking motor turns but the engine does not.
    B. The pinion disengages slowly after starting.
    C. Unusual sounds come from the cranking motor.
    D. The engine cranks slowly but does not start.   (C4)

70. What is the LEAST likely cause of an engine cranking slowly?
    A. A weak battery
    B. Seized pistons or bearings
    C. Low resistance in the starter circuit
    D. High resistance in the starter circuit   (C1)

71. What is the LEAST likely cause of a discharged battery?
    A. A loose alternator belt
    B. A dirty battery cable connection
    C. A defective starter solenoid
    D. A parasitic drain   (B3)

72. Two technicians are discussing an alternator with zero output. Technician A says
    the alternator field circuit may have an open. Technician B says the fusible link
    may be open in the alternator to battery wire. Who is right?
    A. Technician A only
    B. Technician B only
    C. Both A and B
    D. Neither A nor B   (D2)

73. An alternator with a 90-ampere rating produces 45 amps during an output test. A
    V-belt drives the alternator and the belt is at the specified tension. Technician A
    says the V-belt may be worn and bottomed in the pulley. Technician B says the
    alternator pulley may be misaligned with the crankshaft pulley. Who is right?
    A. Technician A only
    B. Technician B only
    C. Both A and B
    D. Neither A nor B   (D3)

74. Technician A says that undercharging could be caused by a loose drive belt.
    Technician B says that undercharging could be caused by undersized wiring
    between the alternator and the battery. Who is right?
    A. Technician A only
    B. Technician B only
    C. Both A and B
    D. Neither A nor B   (D3)

75. Two technicians are discussing a truck charging system output test. Technician A
    says the vehicle accessories should be on during the test. Technician B says the
    charging system voltage should be limited to 17 volts. Who is right?
    A. Technician A only
    B. Technician B only
    C. Both A and B
    D. Neither A nor B   (D4)

76. To perform a voltage drop test on the insulated side of the charging circuit, the
    voltmeter connections should be at:
    A. the regulator output terminal and the field terminal of the alternator.
    B. the battery ground cable and the alternator case.
    C. the output terminal of the alternator and the insulated (positive) terminal of
       the battery.
    D. the field terminal of the alternator and the vehicle frame.   (D5)

77. The ground side of a truck charging circuit is being tested for voltage drop.
    Technician A says to place the voltmeter leads on the voltage regulator ground
    terminal and the vehicle battery. Technician B says to place the voltmeter leads on
    the alternator casing and the truck's battery ground side. Who is right?
    A. Technician A only
    B. Technician B only
    C. Both A and B
    D. Neither A nor B   (D5)

78. Technician A says that a dash-mounted ammeter is the most reliable way to determine whether a charging system is working properly. Technician B states that a dash-mounted voltmeter is a better way to monitor the charging system. Who is right?
    A. Technician A only
    B. Technician B only
    C. Both A and B
    D. Neither A nor B   (D1)

79. What is the best way to test for excessive resistance in the charging circuit?
    A. A visual inspection of the connectors
    B. A voltage drop test with the system at maximum output
    C. Remove the suspect cable(s) and perform a resistance check
    D. Check battery voltage with the engine running   (D5, D7)

80. Technician A says that the battery cables only need to be serviced when the starting or charging system is producing problems. Technician B says that battery corrosion only forms on the terminals during cold weather. Who is right?
    A. Technician A only
    B. Technician B only
    C. Both A and B
    D. Neither A nor B   (C5)

81. Which of these is part of a truck charging system?
    A. Voltage solenoid
    B. Voltage regulator
    C. Voltage transducer
    D. Magnetic switch   (D4)

82. What is the LEAST likely result of a full-fielded alternator?
    A. Low battery voltage due to excessive field current draw
    B. Burned-out light bulbs on the vehicle
    C. A battery that boils over
    D. Excessive charging system voltage   (D8)

83. The vehicle headlights turn off after 30 minutes of operation and then come on after 15 minutes. Technician A says that the headlight switch may be overheating and opening the circuit. Technician B says that a larger circuit breaker should be installed. Who is right?
    A. Technician A only
    B. Technician B only
    C. Both A and B
    D. Neither A nor B   (E1.1)

84. When the driver door of the vehicle is opened, the interior lights illuminate, but very dimly. Technician A says that the headlight switch may be faulty. Technician B says a grounding problem to the door jam switch may be the problem. Who is right?
    A. Technician A only
    B. Technician B only
    C. Both A and B
    D. Neither A nor B   (E1.6)

85. If only the right side of the trailer turn signals are illuminated, the technician should:
    A. replace the turn signal flasher.
    B. replace the bulbs on the left side of the trailer.
    C. check the trailer electrical connection.
    D. check the brake light switch for proper operation.   (E2.2)

86. When diagnosing a repeated flasher failure, technician A says that a 100-millivolt voltage drop problem may be the cause. Technician B says that proper grounding of the trailer sockets may correct the problem. Who is right?
    A. Technician A only
    B. Technician B only
    C. Both A and B
    D. Neither A nor B   (E2.3)

87. All of these are good ways to check the accuracy of an engine temperature sender **EXCEPT:**
    A. check for proper voltage signal output from the sender circuit.
    B. measure the resistance of the sender at a specific temperature.
    C. immerse the sender in boiling water and see if the gauge reads around 212°F (100°C).
    D. check the reading of the engine temperature sender against that of other senders in the truck after it has been sitting overnight.   (F3)

6 Additional Test Questions for Practice

Additional Test Questions

Please note the letter and number in parentheses following each sample question. They match the overview in section 4 that discusses the relevant subject matter. You may want to refer to the overview using this cross-referencing key to help with questions posing problems for you.

1. Technician A says that a starter drive pinion should not have chamfers on the drive teeth. Technician B states that if the flywheel ring gear is damaged, then the entire flywheel should be replaced. Who is correct?
   A. Technician A only
   B. Technician B only
   C. Both A and B
   D. Neither A nor B
   (C4)

2. Technician A says that overcharging a battery will not cause significant long-term damage. Technician B says that in hot weather, more current is needed to charge a battery. Who is right?
   A. Technician A only
   B. Technician B only
   C. Both A and B
   D. Neither A nor B
   (B5)

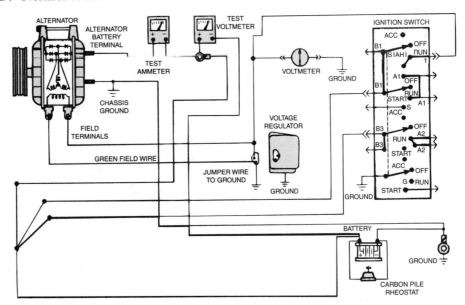

3. In the charging system shown above, the voltmeter reading will show:
   A. charging system voltage.
   B. regulator operating voltage.
   C. charging system voltage drop.
   D. ignition switch voltage drop.
   (D5)

4. Two technicians are discussing battery terminal connections. Technician A says when disconnecting battery cables, always disconnect the negative cable first. Technician B says when connecting battery cables, always connect the negative cable first. Who is right?
   A. Technician A only
   B. Technician B only
   C. Both A and B
   D. Neither A nor B   (B3)

5. Technician A says that a charge indicator light should come on with the key on, engine off. Technician B states that if the indicator is not on with the engine running, this means that the charging system is functioning properly. Who is right?
   A. Technician A only
   B. Technician B only
   C. Both A and B
   D. Neither A nor B   (D1)

6. On a truck with an indicator light the charge light is not on when the engine is running and the truck has an undercharged battery. Technician A says this can be caused by a blown fuse between the indicator lamp and ignition switch. Technician B says this could be caused by a burned-out bulb. Who is right?
   A. Technician A only
   B. Technician B only
   C. Both A and B
   D. Neither A nor B   (D1)

7. Two technicians are discussing a dash-mounted voltmeter. Technician A says that it is normal for the dash voltmeter to read a different voltage from that of a test voltmeter across the battery terminals. Technician B says that resistance in the dash voltmeter ground will affect its readings. Who is right?
   A. Technician A only
   B. Technician B only
   C. Both A and B
   D. Neither A nor B   (D1)

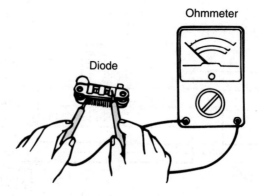

8. Technician A says a diode check is being performed in the figure shown above. Technician B says a voltage drop test is being performed in the figure. Who is right?
   A. Technician A only
   B. Technician B only
   C. Both A and B only
   D. Neither A nor B   (A8)

9. You are working on a truck and find that a battery cable terminal end is badly corroded. All of the following are proper repair procedures **EXCEPT:**
   A. replacing the entire cable assembly.
   B. replacing the terminal with a bolt-on end and heat shrink tubing.
   C. replacing the terminal with a crimp-on end and heat shrink tubing.
   D. replacing the terminal with a soldered end and heat shrink tubing.   (C5)

10. When performing an alternator maximum output test, what is the most practical and safe way to make the alternator put out maximum output?
   A. Using a carbon pile across the battery
   B. Full-fielding the alternator
   C. Turning on all of the vehicle electrical loads
   D. Temporarily installing a discharged battery into the vehicle   (D4)

11. Technician A says that when performing an alternator output test, a voltmeter should be connected in series with the alternator output terminal and the battery ground cable. Technician B says that a carbon pile should be used when performing an alternator output test. Who is right?
   A. Technician A only
   B. Technician B only
   C. Both A and B
   D. Neither A nor B   (D4)

12. An alternator output test is being performed. Technician A uses only a voltmeter connected across the battery positive terminal and negative terminal while the engine is running. Technician B says a carbon pile is not needed since the engine is already running. Who is right?
   A. Technician A only
   B. Technician B only
   C. Both A and B
   D. Neither A nor B   (D4)

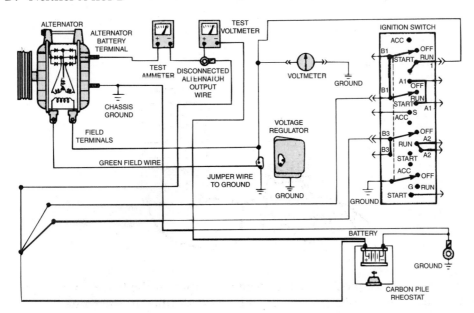

13. With an ammeter and voltmeter connected to the charging system as shown in the above figure, the voltmeter indicates 2 volts and the ammeter reads 100 amps. Technician A says this condition may cause an undercharged battery. Technician B says this condition may result in a headlight flare-up during acceleration. Who is right?
   A. Technician A only
   B. Technician B only
   C. Both A and B
   D. Neither A nor B   (D5)

14. What is the best way to determine if a problem exists with an alternator output wire?
    A. Visual inspection
    B. Remove the wire and check its resistance with an ohmmeter
    C. Perform a voltage drop test through it with the engine off
    D. Perform a voltage drop test through it with the alternator at maximum output
    (D7)

15. When removing an alternator, Technician A says that you should remove the positive battery cable first. Technician B says that all of the wires on the back of an alternator are polarized and that you need not worry about labeling them before removal. Who is right?
    A. Technician A only
    B. Technician B only
    C. Both A and B
    D. Neither A nor B
    (D6)

16. You are testing alternator output. Immediately after starting the engine, but before loading the battery, you find that the current output from the alternator slowly decreases the longer the engine runs. What can this mean?
    A. Alternator output is marginal; discontinue the test.
    B. The drive belt is most likely slipping on the pulley.
    C. The battery is slowly recovering to capacity.
    D. The diodes in the alternator are heating up and starting to fail.
    (D4, B5)

17. What is the LEAST likely result of a full-fielded alternator?
    A. High battery voltage level
    B. Battery gassing
    C. Low battery voltage level
    D. High alternator amperage output
    (D8)

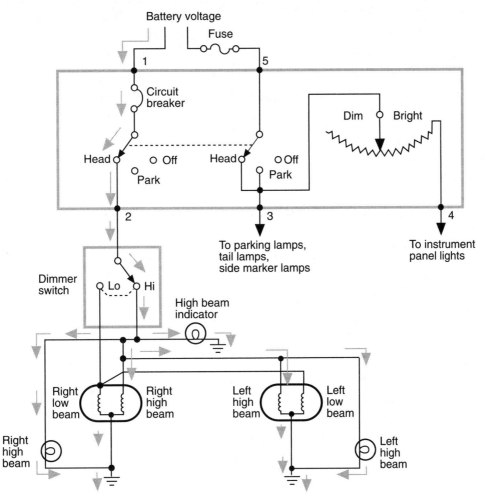

18. The left-side headlight is dim only on the high beam in the above figure. The other headlights operate normally. Technician A says there may be high resistance in the left-side headlight ground. Technician B says there may be high resistance in the dimmer switch high beam contacts. Who is right?
    A. Technician A only
    B. Technician B only
    C. Both A and B
    D. Neither A nor B   (E1.1)

19. Technician A says that headlight aim should be checked on a level floor with the vehicle unloaded. Technician B says that this may conflict with existing laws and regulations in some states. Who is right?
    A. Technician A only
    B. Technician B only
    C. Both A and B
    D. Neither A nor B   (E1.2)

20. A headlight aiming procedure is being performed. Technician A says that when headlight aiming equipment is not available, headlight aiming can be checked by projecting the high beam of each light upon a screen or chart at a distance of 25 feet ahead of the headlights. Technician B says that when aiming, the vehicle should be exactly parallel to the chart or screen. Who is right?
    A. Technician A only
    B. Technician B only
    C. Both A and B
    D. Neither A nor B   (E1.2)

21. Which of these is the LEAST likely cause of a dim headlight?
    A. Corrosion on the headlight connector
    B. Slightly damaged or broken headlamp assembly
    C. Alternator output low
    D. High resistance in the wiring to the headlamp assembly   (E1.1)

22. Technician A says that when checking an inoperative component, it should be checked downstream from that component for an open. Technician B says to use an ammeter when checking for continuity in a circuit. Who is right?
    A. Technician A only
    B. Technician B only
    C. Both A and B
    D. Neither A nor B   (A5)

23. A truck/trailer comes in with a dim left-rear taillight on the trailer. Technician A says that this may be due to a corroded pin on the left side of the trailer connector. Technician B states that the problem may be caused by a poor ground connection at the left-rear taillight. Who is right?
    A. Technician A only
    B. Technician B only
    C. Both A and B
    D. Neither A nor B   (A5)

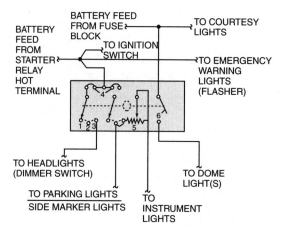

24. In the figure shown above, the headlight switch is in what position?
    A. Park
    B. Run
    C. On
    D. Off   (E1.3)

25. Technician A says that it is a good idea to coat the prongs and base of a new sealed beam with dielectric grease before installing to prevent possible corrosion. Technician B says that white lithium grease can be used instead of dielectric grease. Who is right?
    A. Technician A only
    B. Technician B only
    C. Both A and B
    D. Neither A nor B   (E1.2)

26. Which of the following components is used to dim dash lights?
    A. Voltage limiter
    B. Rheostat
    C. Resistor
    D. Diode   (E1.5)

27. Technician A says that an alternator that is overcharging usually causes lights that are brighter than normal. Technician B states that poor chassis grounds usually cause dim lights. Who is correct?
    A. Technician A only
    B. Technician B only
    C. Both A and B
    D. Neither A nor B   (E1.1)

28. Technician A says that a headlight dimmer switch can be floor mounted. Technician B says that a headlight dimmer switch can be mounted on the front panel. Who is right?
    A. Technician A only
    B. Technician B only
    C. Both A and B
    D. Neither A nor B   (E1.3)

29. A truck is having an intermittent fault with its high beams. All of these could be a possible cause **EXCEPT:**
    A. a defective headlight dimmer switch.
    B. defective high beam filaments inside headlamps.
    C. a loose wiring harness connector.
    D. a defective headlight switch.   (E1.3)

30. A problem is encountered with the interior cab lights of a truck. Every time that a courtesy light is replaced, the fuse protecting that circuit blows. Technician A says that this could be caused by a short upstream (toward source voltage) from the light. Technician B says that this could be caused by a short downstream (toward circuit ground) from the respective light. Who is right?
    A. Technician A only
    B. Technician B only
    C. Both A and B
    D. Neither A nor B   (E1.6)

31. Technician A says that a multifunction switch may have a cruise control function. Technician B uses the multifunction switch as a turn signal switch. Who is right?
    A. Technician A only
    B. Technician B only
    C. Both A and B
    D. Neither A nor B   (E2.2)

32. A trailer has inoperative taillights on one side only. Technician A says to check the trailer circuit connector for an open. Technician B uses an ohmmeter to check the continuity between the defective side and the trailer circuit connector with the circuit under power. Who is right?
    A. Technician A only
    B. Technician B only
    C. Both A and B
    D. Neither A nor B   (E1.4)

33. When replacing halogen headlight bulbs, Technician A always wears gloves. Technician B says that halogen bulbs outlast conventional sealed-beam headlights. Who is right?
    A. Technician A only
    B. Technician B only
    C. Both A and B
    D. Neither A nor B   (E1.2)

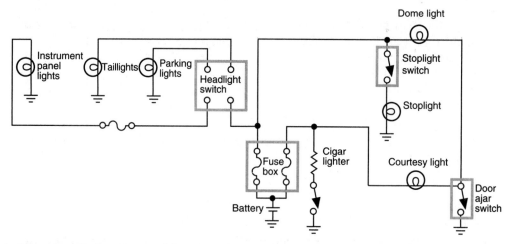

34. The cigar lighter fuse is blown in the stoplight circuit in the figure above. The result of this problem is most likely:
    A. the courtesy and dome lights come on dimly when you push in the lighter.
    B. the stop and dome lights are completely inoperative.
    C. the parking lights, taillights, and instrument panel lights are inoperative.
    D. the courtesy light will not work.

    (E2.1)

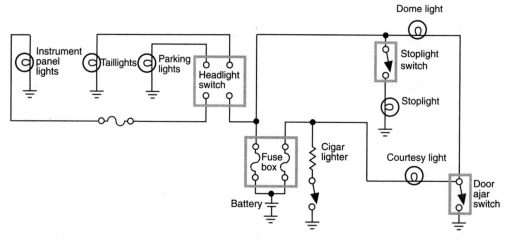

35. In the figure above, the headlight switch has failed. Technician A says the parking lights are not affected. Technician B says that the stoplights will fail to operate. Who is right?
    A. Technician A only
    B. Technician B only
    C. Both A and B
    D. Neither A nor B

    (E2.1)

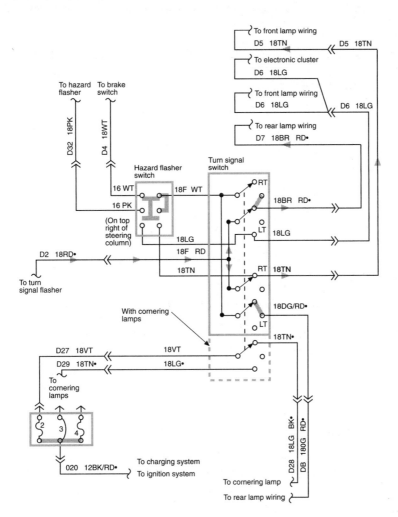

36. In the turn signal light circuit shown above, the right rear signal light is dim, and the other lights work normally. Which of these is most likely the cause?
    A. High resistance in DB 180G RD from the turn signal light switch to the rear lamp wiring
    B. A short to ground in the DB 180G RD wire from the turn signal light switch to the rear lamp wiring
    C. High resistance in D7 18BR RD from the turn signal light switch to the rear lamp wiring
    D. High resistance in D2 18 RD from the turn signal flasher to the rear lamp wiring   (E2.2)

37. A truck has a turn signal complaint. You find that the left front light does not work and the left rear light flashes slower than normal. Technician A says the left front bulb is faulty. Technician B says there could be an open circuit between the switch and the left front bulb. Who is right?
    A. Technician A only
    B. Technician B only
    C. Both A and B
    D. Neither A nor B   (E2.3)

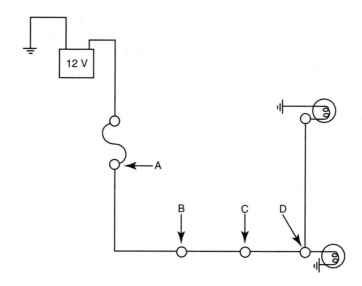

38. In the figure shown above, the backup light circuit fuse keeps blowing. Technician A uses an ohmmeter to check the circuit between points A and B. Technician B uses an ohmmeter to check the circuit between points C and ground. Who is right?
    A. Technician A only
    B. Technician B only
    C. Both A and B
    D. Neither A nor B   (E2.4)

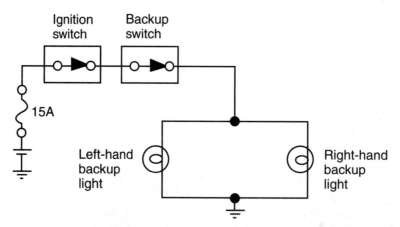

39. The right-hand backup light circuit is accidentally grounded on the switch side of the bulb in the circuit shown above. Technician A says this condition may blow the backup light fuse. Technician B says the left-hand backup light may work normally while the right-hand backup light is inoperative. Who is right?
    A. Technician A only
    B. Technician B only
    C. Both A and B
    D. Neither A nor B   (E2.4)

40. Technician A says that all stoplight switches are air activated. Technician B states that stoplight switches send current directly to the stoplights. Who is right?
    A. Technician A only
    B. Technician B only
    C. Both A and B
    D. Neither A nor B   (E2.1)

41. The dash light on a medium truck does not work. Technician A says that the fuse to the taillights could be the cause. Technician B says that the rheostat in the headlight switch could be the cause. Who is right?
    A. Technician A only
    B. Technician B only
    C. Both A and B
    D. Neither A nor B   (E1.5)

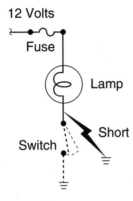

42. The interior cab light circuit in the above figure in a truck has a short to ground. Technician A says current flow through the lamp is higher than normal. Technician B says the light cannot be turned off. Who is right?
    A. Technician A only
    B. Technician B only
    C. Both A and B
    D. Neither A nor B   (E1.6)

43. Technician A says that older style bi-metal gauges are not sensitive to voltage fluctuations in the electrical system. Technician B says that most magnetic type gauges are not sensitive to voltage changes in the electrical system. Who is right?
    A. Technician A only
    B. Technician B only
    C. Both A and B
    D. Neither A nor B   (F1)

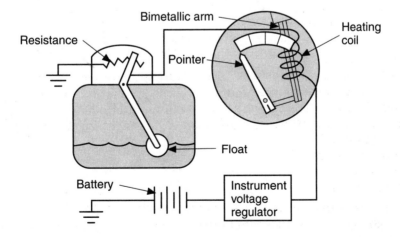

44. The fuel gauge in the figure above reads lower than the actual fuel tank level. All of the other gauges operate normally. Which of these is the possible cause?
    A. High resistance in the sending unit ground
    B. High resistance after the voltage regulator
    C. A short to ground between the gauge and the sending unit
    D. An open circuit in the wire from the gauge to the sending unit   (F1)

45. All the gauges are erratic in an instrument panel with thermal-electric gauges and an instrument voltage limiter. Technician A says the alternator may be at fault. Technician B says the instrument voltage limiter may be defective. Who is right?
    A. Technician A only
    B. Technician B only
    C. Both A and B
    D. Neither A nor B   (F1)

46. If a voltage limiter fails, it could cause:
    A. all gauges to read high.
    B. the temperature gauge to read low.
    C. all gauges to read low
    D. erratic operation of one or all gauges.   (F1)

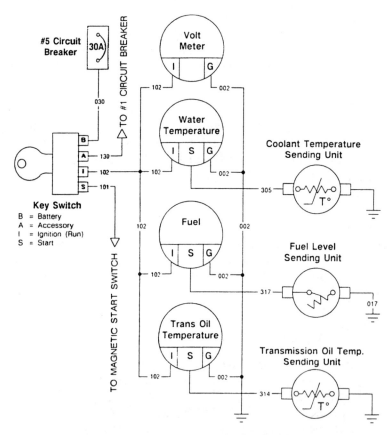

47. In the figure shown above, the ground circuit for the fuel sender (#017) has failed. Technician A says that only the fuel gauge will be affected. Technician B states that all of the gauges will be affected because the gauges share a common ground. Who is right?
    A. Technician A only
    B. Technician B only
    C. Both A and B
    D. Neither A nor B   (F3)

48. Technician A says that a truck with an electronic instrument panel sources its information directly from the sensors on the engine. Technician B states that a data bus style electronic instrument panel receives information from the engine ECM. Who is right?
    A. Technician A only
    B. Technician B only
    C. Both A and B
    D. Neither A nor B   (F2)

49. On a truck with electronic instrumentation, gauge accuracy is suspect. What would be the best way to determine where the fault lies?
    A. Swap the panel with a like-new one
    B. Replace the suspect sender unit(s)
    C. Ground the sender wire at the suspect sender unit(s)
    D. Check the values that the ECM is displaying using a digital diagnostic reader and compare to gauge readings   (F2)

50. Warning lights and warning devices are generally activated by:
    A. the vehicle ignition switch.
    B. closing of a switch or sensor.
    C. opening of a switch or sensor.
    D. the vehicle battery.   (F4)

51. The speedometer in an electronically managed truck is not accurate. Which of the following is the LEAST likely problem?
    A. The rear axle ratio has not been correctly entered into the engine computer.
    B. The transmission speed sensor has not been calibrated.
    C. The tire rolling radius has not been correctly entered into the engine computer.
    D. The engine computer was not reprogrammed when new rear tires were installed.   (F5)

52. Technician A says that electric gauges can be checked by removing the wire from the sender and grounding it to the block. Technician B states that a good way to test electric gauges for accuracy is to substitute the sender value using a variable resistance test box. Who is right?
    A. Technician A only
    B. Technician B only
    C. Both A and B
    D. Neither A nor B   (F3)

53. What does the pyrometer do?
    A. Monitors fuel flow
    B. Indicates the battery state of charge
    C. Indicates engine speed
    D. Monitors temperature changes   (F3)

54. When checking an electric water temperature gauge for an erratic reading, technician A uses an ohmmeter to check gauge resistance. Technician B places the sensing bulb in boiling water and compares the gauge reading against that of a thermometer placed in the water. Who is right?
    A. Technician A only
    B. Technician B only
    C. Both A and B
    D. Neither A nor B   (F3)

55. The "check engine" light illuminates while a truck is being operated. What does this indicate?
    A. Low engine oil pressure
    B. High engine coolant temperature
    C. Maintenance reminder
    D. Low coolant level   (F4)

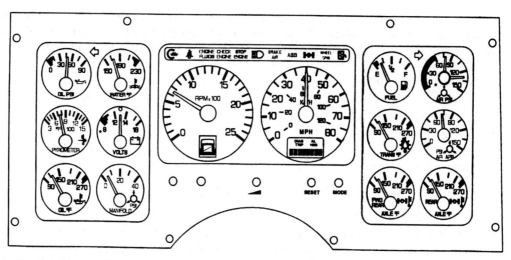

56. In the figure shown above of a typical truck instrument panel, which set of gauges would likely source their information from the engine ECM?
    A. All of them
    B. Only the middle and right-hand clusters
    C. Only the left- and right-hand clusters
    D. Only the left-hand and middle clusters   (F2)

57. On a truck with data bus driven gauges, all of the gauge needles sweep from left to right immediately after turning the key on. Technician A says that this may indicate a fault with the instrument panel. Technician B states that this is due to battery voltage that is too high. Who is right?
    A. Technician A only
    B. Technician B only
    C. Both A and B
    D. Neither A nor B   (F2)

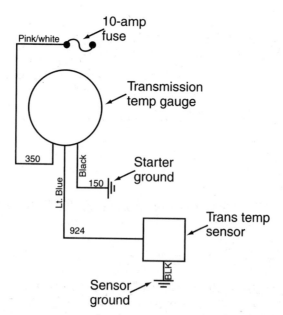

58. In the figure shown above, what would happen if circuit 924 were open?
    A. The gauge would read high.
    B. The gauge would read low.
    C. The gauge would fluctuate.
    D. The gauge would be inoperative.   (F3)

59. Technician A says that the vehicle gauges are accurate enough for diagnosing most engine problems. Technician B states that questionable readings should always be confirmed with another gauge before major repair decisions are made. Who is right?
    A. Technician A only
    B. Technician B only
    C. Both A and B
    D. Neither A nor B   (F1)

60. All of the following could cause an inaccurate gauge reading **EXCEPT:**
    A. a defective ground at the sender unit.
    B. high battery voltage.
    C. a defective IVR (instrument voltage regulator).
    D. excessive resistance in the gauge wiring.   (F3)

61. Technician A says it is acceptable to use test lights on electronic circuits and systems such as the instrument cluster. Technician B says test lights are OK for electrical circuits such as headlamps, horns, and power accessories but that a digital multimeter (DMM) should be used on electronic control circuits. Who is right?
    A. Technician A only
    B. Technician B only
    C. Both A and B
    D. Neither A nor B   (F4)

62. All of the following are ways to signal an electronic tachometer **EXCEPT:**
    A. using information from the data bus.
    B. from the alternator phase tap, or "R" terminal.
    C. from an injector driver unit.
    D. from a remote mounted signal generator driven by the engine.   (F5)

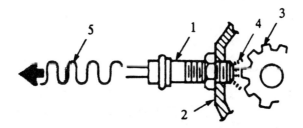

63. The type of sensor shown in the above figure can be used for what?
    A. Speed sensing
    B. Pressure sensing
    C. Temperature sensing
    D. Level sensing   (F5)

64. The range of error allowed in a test of a dual pressure air gauge is:
    A. 2 psi.
    B. 8 psi.
    C. 4 psi.
    D. 10 psi.   (F1)

65. The component that provides a tach signal to the instrument cluster on older medium-duty trucks with diesel engines is:
    A. primary ignition coil.
    B. secondary ignition coil.
    C. generator (R terminal).
    D. vehicle speed sensor.   (F5)

66. A customer says the horn on a truck will not turn off. Technician A says the cause could be welded diaphragm contacts inside the horn. Technician B says the relay may be defective. Who is right?
    A. Technician A only
    B. Technician B only
    C. Both A and B
    D. Neither A nor B

(G1)

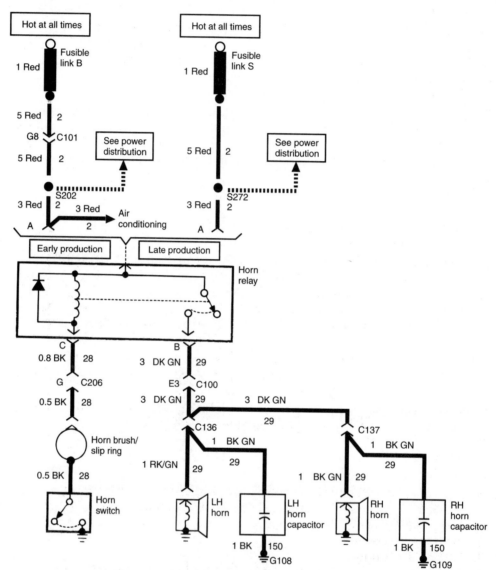

67. In the circuit shown above there is no horn operation. All of these may be the cause of the problem **EXCEPT**:
    A. an open ground circuit in the horn relay.
    B. an open circuit in the horn relay winding.
    C. an open circuit at the horn brush/slip ring.
    D. an open fusible link in the relay power wire.

(G1)

68. Using the figure shown for question #67, the horn blows continuously. Technician A says that a short to ground at connector C206 may cause this. Technician B states that a short to ground at connector C100 could cause this problem. Who is right?
    A. Technician A only
    B. Technician B only
    C. Both A and B
    D. Neither A nor B   (G1)

69. A vehicle electric horn does not work when the horn is depressed. Technician A uses a test lamp to check the power and ground terminals of the horn relay. Technician B uses a DMM to check the power and ground terminals of the horn relay. Who is right?
    A. Technician A only
    B. Technician B only
    C. Both A and B
    D. Neither A nor B   (G2)

70. A horn on a medium-duty truck operates intermittently; which of these could be the cause?
    A. Blown fuse
    B. No power to relay
    C. Open in horn button circuit
    D. Defective horn relay   (G2)

71. A wiper motor fails to operate. Which of the following would be the LEAST likely cause?
    A. A defective wiper switch
    B. A tripped thermal overload protector
    C. A tripped circuit breaker
    D. Resistance in the motor supply wiring   (G3)

72. A 2-speed wiper motor is being discussed. Technician A says that 2-speed operation is enabled using separate sets of high and low speed brushes inside the motor. Technician B states that the 2-speed operation is accomplished by using an external resistor pack similar to a heater blower motor. Who is right?
    A. Technician A only
    B. Technician B only
    C. Both A and B
    D. Neither A nor B   (G3)

73. A heavy truck windshield wiper system is being inspected. Technician A says that air or electricity drives the wipers. Technician B says that one or two motors operate a wiper system. Who is right?
    A. Technician A only
    B. Technician B only
    C. Both A and B
    D. Neither A nor B   (G3)

74. A wiper motor operates very sluggishly. Technician A says that this might be due to poor brush contacts inside the motor. Technician B states that this might be caused by an open in the motor ground circuit. Who is right?
    A. Technician A only
    B. Technician B only
    C. Both A and B
    D. Neither A nor B   (G3)

75. What is the function of the park circuit in the windshield wiper system?
    A. Stop the wiper blades at the same position regardless of switch position
    B. Shut down the wiper motor in case of overheating
    C. Help keep the wiper blades synchronized
    D. Stop the wiper motor in case of a low voltage problem   (G4)

76. If the compressor clutch diode fails, which of these is the resultant condition?
    A. ECM and CPU failure from voltage spikes
    B. The compressor runs backward.
    C. Clutch coil inoperative due to no current
    D. Compressor clutch coil failure from high current   (A8)

77. The wipers on a truck equipped with electric windshield wipers will not park.
    Technician A says the activation arm is broken or out of adjustment. Technician B
    says a faulty wiper switch will cause this condition. Who is right?
    A. Technician A only
    B. Technician B only
    C. Both A and B
    D. Neither A nor B   (G4)

78. Technician A says binding mechanical wiper linkage can cause no wiper operation.
    Technician B says a shorted control circuit can cause constant wiper operation.
    Who is right?
    A. Technician A only
    B. Technician B only
    C. Both A and B
    D. Neither A nor B   (G5)

79. A technician is diagnosing a complaint of a windshield wiper motor that turns
    very slowly. He suspects a problem with high resistance in the wiring between the
    switch and the motor. What should be his course of action?
    A. Disconnect the wire and measure the resistance through it.
    B. Install an ammeter in series and measure the current flow through it.
    C. Perform a voltage drop test along that length of wire.
    D. Measure the operating voltage at the motor end of the wire.   (A2, G4)

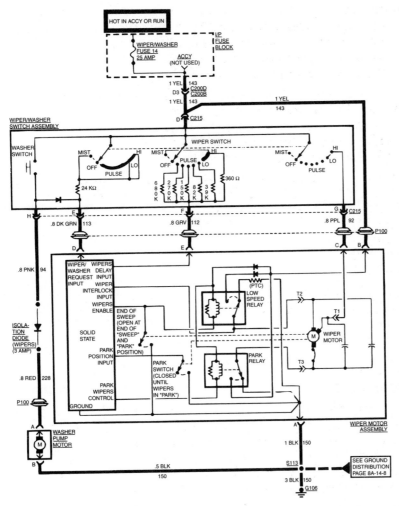

80. In the circuit shown above, the windshield washer does not operate. The wiper motor operates normally. Technician A says the wiper/washer fuse may be open. Technician B says the isolation diode may have an open circuit. Who is right?
    A. Technician A only
    B. Technician B only
    C. Both A and B
    D. Neither A nor B   (G6)

81. Using the figure from question #80, the wiper washer pump motor runs constantly. What might be the problem?
    A. The ground side of the motor is shorted to ground.
    B. The control switch is shorted to ground.
    C. The contacts in the switch are stuck closed.
    D. The wiper washer pump relay contacts are stuck closed.   (G6)

82. A heated mirror is being checked for a complaint related to constant fuse blowing. A current draw check with an ammeter shows that the heater unit draws a varying amount of amperage. Technician A says that this may be due to the fact that the resistance of the heater changes as it warms up. Technician B states that this condition is normal. Who is right?
    A. Technician A only
    B. Technician B only
    C. Both A and B
    D. Neither A nor B   (G7)

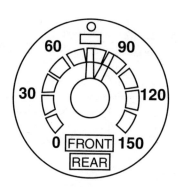

83. Technician A says that the component in the figure shown above could be a dual ammeter. Technician B says that the component in the figure shown above could be a dual air pressure gauge. Who is right?
    A. Technician A only
    B. Technician B only
    C. Both A and B
    D. Neither A nor B   (F1)

84. The state of charge of a battery that has a specific gravity of 1.20 at 80°F would be:
    A. completely discharged.
    B. about 3/4 charged.
    C. about 1/2 charged.
    D. fully charged.   (B2)

85. An intermittent problem with inoperative heated mirrors is encountered. Technician A checks for a reliable power supply first. Technician B says that it is a good idea to check the wiring leading to mirrors when an intermittent problem is suspected. Who is right?
    A. Technician A only
    B. Technician B only
    C. Both A and B
    D. Neither A nor B   (G7)

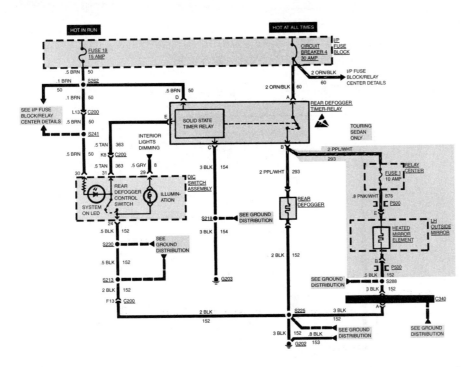

86. In the circuit shown above, the heated mirror element does not function, but the rear defogger operates normally. All of the following could cause this problem **EXCEPT:**
    A. an open circuit in the number 4 circuit breaker in the fuse block.
    B. a blown number 1 fuse in the relay center.
    C. an open circuit between the heated mirror element and ground.
    D. an open circuit between the timer relay and the number 1 fuse.   (G7)

87. When checking the power mirror and heated mirror circuit, technician A says to replace the circuit breaker whenever the system experiences problems. Technician B says to change the fuse whenever the system experiences problems. Who is right?
    A. Technician A only
    B. Technician B only
    C. Both A and B
    D. Neither A nor B   (G7)

88. Technician A says that if a heater blower motor resistor burns out, the unit will still operate on high speed. Technician B states that if a heater blower motor resistor burns out, it will not operate because the resistors are wired in series. Who is right?
    A. Technician A
    B. Technician B
    C. Both A and B
    D. Neither A nor B   (G8)

89. In a medium-duty truck, the largest fusible link is usually located:
    A. near the ignition switch.
    B. near the vehicle PCM (powertrain control module).
    C. in the instrument panel.
    D. at the starter solenoid battery terminal.   (A7)

90. The function of a fusible link is to:
    A. take the place of a circuit breaker.
    B. break in half during a current overload.
    C. eliminate fuses.
    D. open an overloaded circuit while maintaining insulation protection.   (A7)

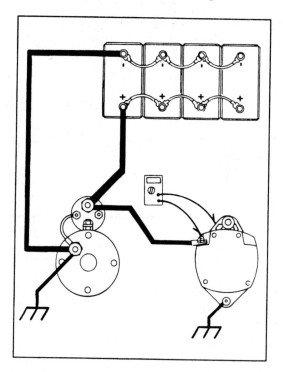

91. Referencing the figure shown above, what test is being performed?
    A. Voltage output test
    B. Positive charging circuit cable voltage drop test
    C. Charging ground circuit voltage drop test
    D. Starter operating voltage test   (A2)

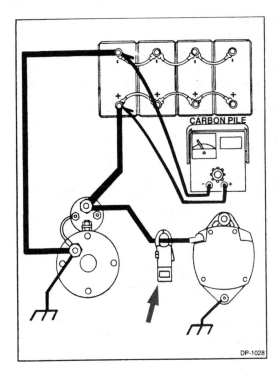

DP-1028

92. Using the figure shown above, what test is being performed with the unit denoted by the arrow?
   A. Starter current draw test
   B. Battery load test
   C. Alternator output test
   D. Parasitic battery draw test   (D4)

93. While reassembling an alternator during a rebuild, technician A replaces the pulley drive. Technician B says that it is good practice to replace the field brushes when performing an alternator rebuild. Who is right?
   A. Technician A only
   B. Technician B only
   C. Both A and B
   D. Neither A nor B   (D6)

94. A test of the starter control circuit may reveal problems in which area?
   A. Battery cable connections
   B. Starter motor
   C. Starter solenoid high-current contacts
   D. Magnetic switch   (C2)

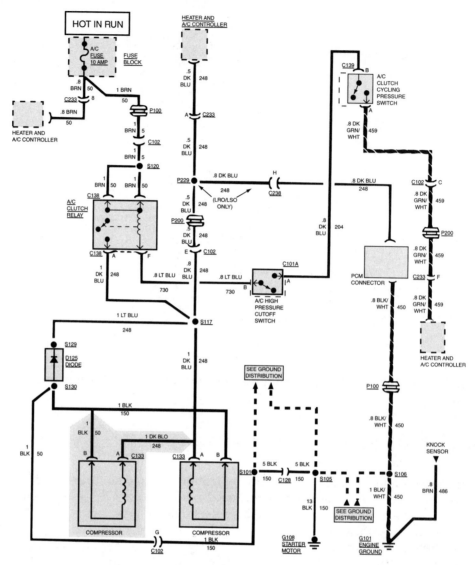

95. An electrical schematic is being examined in the figure shown above. Technician A says that this is the heater and A/C controller circuit. Technician B says that this circuit operates with circuit #50 (1 brn) being open. Who is right?
    A. Technician A only
    B. Technician B only
    C. Both A and B
    D. Neither A nor B

                                                                          (G8)

96. Technician A says that most blower motors used in medium and heavy trucks use AC voltage for power. Technician B states that some blower motor circuits incorporate a relay between the resistor pack and the motor to handle the high-current requirements. Who is right?
    A. Technician A only
    B. Technician B only
    C. Both A and B
    D. Neither A nor B

                                                                          (G8)

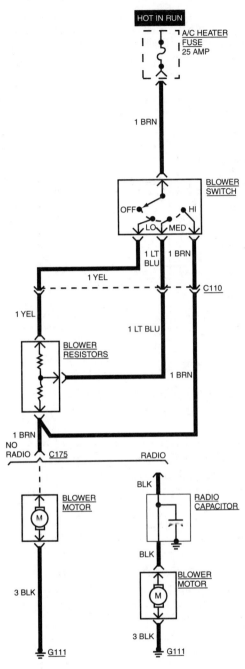

97. In the figure above, when the blower resistors are removed, the blower motor will:
    A. not operate.
    B. blow the system fuse.
    C. operate on high speed only.
    D. operate on low speed only.   (G8)

98. Technician A uses a test lamp to detect resistance. Technician B uses a jumper wire to test circuit breakers, relays, and lights. Who is right?
    A. Technician A only
    B. Technician B only
    C. Both A and B
    D. Neither A nor B   (A4)

99. Technician A says that a voltage drop of 500 millivolts per connection is acceptable. Technician B says that a voltage drop of 200 millivolts per connection is acceptable. Who is right?
    A. Technician A only
    B. Technician B only
    C. Both A and B
    D. Neither A nor B

(A2)

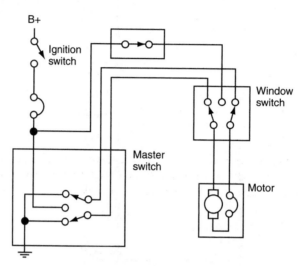

100. In the figure above, the power window operates normally from the master switch, but the window does not work using the window switch. Which of the following could be the cause?
    A. An open between the ignition switch and window switch
    B. An open in the window switch movable contacts
    C. An open in the master switch ground wire
    D. A short to ground at the circuit breaker in the motor

(G10)

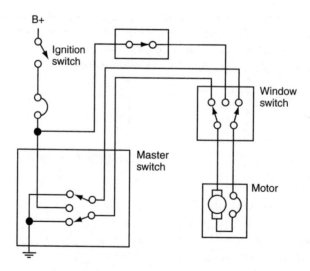

101. In the figure above, the vehicle power window circuit is inoperative. The technician should first:
    A. install a jumper wire across the circuit breaker or fuse.
    B. remove the door panels and apply voltage to the window motors.
    C. check for power at the circuit breaker or fuse.
    D. measure the resistance of the window motors.

(G10)

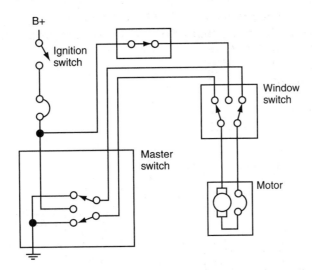

102. In the figure above, the windows will not operate from either the master or door switches. Technician A says that a defective diode in the circuit could be the cause. Technician B says that corrosion at the window switch contacts may be the cause. Who is right?
     A. Technician A only
     B. Technician B only
     C. Both A and B
     D. Neither A nor B   (G10)

103. A truck is examined for inoperative power windows. Technician A says to check the passenger side first by pulling off the door panel and checking the motor. Technician B checks alternator output first to check for proper voltage. Who is right?
     A. Technician A only
     B. Technician B only
     C. Both A and B
     D. Neither A nor B   (G11)

104. A truck with electronic engine controls is being discussed. Technician A says that the fan may be engaged by either positive or ground side switching of the fan relay. Technician B states that a fan will only be driven when the operating temperatures exceed programmed limits. Who is right?
     A. Technician A only
     B. Technician B only
     C. Both A and B
     D. Neither A nor   (G13)

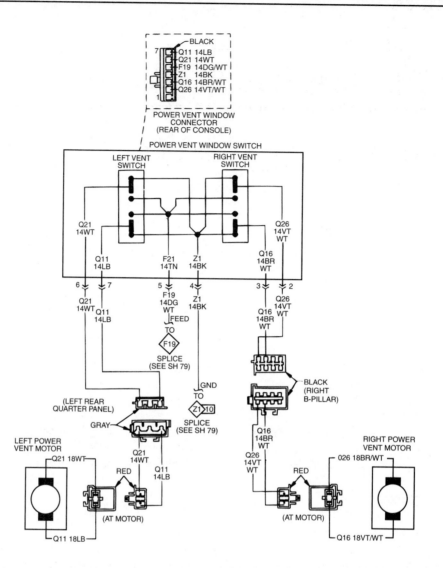

105. In the circuit shown above, the right power vent window does not operate. The left power vent window operates normally. Technician A says there may be an open between the right power window switch and ground. Technician B says the upper contacts in the right power vent window switch may be open. Who is right?
    A. Technician A only
    B. Technician B only
    C. Both A and B
    D. Neither A nor B   (G11)

106. A truck electronic cruise control will not operate, although everything else functions properly. Which of the following should a technician check first?
    A. Servo motor
    B. Control switches in the cab
    C. Linkage to the fuel pump
    D. Vehicle speed   (G12)

107. A truck with an electronically controlled engine comes in with its fan constantly engaged. Technician A says that this may be due to a defective temperature sensor somewhere on the engine. Technician B states that this could be caused by continually operating the truck with the cruise control on. Who is right?
    A. Technician A only
    B. Technician B only
    C. Both A and B
    D. Neither A nor B   (G13)

Appendices

Answers to the Test Questions for the Sample Test Section 5

1.	C	24.	B	47.	C	70.	C
2.	A	25.	A	48.	B	71.	C
3.	B	26.	D	49.	D	72.	C
4.	D	27.	D	50.	D	73.	A
5.	C	28.	A	51.	B	74.	C
6.	A	29.	A	52.	C	75.	D
7.	D	30.	C	53.	A	76.	C
8.	D	31.	A	54.	B	77.	B
9.	B	32.	D	55.	A	78.	B
10.	C	33.	B	56.	A	79.	B
11.	B	34.	D	57.	A	80.	D
12.	D	35.	C	58.	C	81.	B
13.	A	36.	A	59.	D	82.	A
14.	C	37.	B	60.	D	83.	A
15.	C	38.	B	61.	A	84.	B
16.	B	39.	D	62.	C	85.	C
17.	B	40.	A	63.	A	86.	D
18.	A	41.	C	64.	C	87.	A
19.	D	42.	C	65.	B		
20.	D	43.	C	66.	C		
21.	D	44.	C	67.	C		
22.	A	45.	B	68.	A		
23.	A	46.	A	69.	D		

Explanations to the Answers for the Sample Test Section 5

Question #1
Answer A is wrong because connecting the test lamp ground clip at the fuse will not properly ground the test light. The fuse is the battery positive side of this circuit.
Answer B is wrong because the battery positive terminal is also the positive side of this circuit. The ground clip needs to be connected to a battery or chassis ground.
Answer C is correct because the battery negative terminal is a good place to clip the ground side of the test light to allow it to function properly.
Answer D is wrong because the battery positive side of the switch is also on the positive side of this circuit. The ground clip of the test light needs to be connected to a good battery or chassis ground in order to operate properly.

Question #2
Answer A is correct because the picture shows the circuit open at the connector. It has been pulled apart for testing purposes.
Answer B is wrong because an ohmmeter should never be used on a live power circuit. Strange readings and possible meter damage will result.
Answer C is wrong because only technician A is correct.
Answer D is wrong because one of the technicians is correct.

Question #3
Answer A is wrong because an ohmmeter can be used to check a circuit for continuity. The ohmmeter uses internal power to circulate a small amount of current through the circuit to check resistance and ultimately continuity.
Answer B is correct because although a person could use an ammeter to check for continuity, it is very impractical to do so. The other three choices are much better tools for checking continuity. An ammeter is used primarily to check for the amount of current flow.
Answer C is wrong because a voltmeter can be used to probe at various points along a circuit to check that voltage exists sequentially up to the load. It can also be used to check for continuity on the ground side of a load.
Answer D is wrong because a test lamp is also an excellent and easy way to check for continuity. It is used much like the voltmeter in actual operation.

Question #4
Answer A is wrong because a DMM cannot generate anywhere near 30 amps of current when using the ohmmeter function. It will only circulate enough current through a circuit to determine its resistance.
Answer B is wrong because if a circuit breaker or fuse is good, the ohmmeter should show a very low reading (near zero), not infinity, which would indicate an open circuit.
Answer C is wrong because neither technician is correct.
Answer D is correct because both technicians are wrong.

Question #5
Answer A is a good choice, because only a DMM is an appropriate tool to use for testing voltages in an electronic circuit. However, since technician B is also correct, answer A is wrong.
Answer B is also a good choice, because a digital multimeter should have a minimum of 10 megohms impedance to prevent it from significantly altering a circuit's measurable parameters. However, since technician A is also correct, answer B is wrong.
Answer C is correct because both technicians are right.
Answer D is wrong because both technicians are correct.

Question #6
Answer A is correct because a blown or open fuse will cause an open circuit, which would give a zero volt reading on a voltmeter.
Answer B is wrong because a defective bulb would not prevent a voltage reading to the positive side of the voltmeter. Nor would it affect the ground side of the voltmeter.
Answer C is wrong because if the bulb was OK, battery voltage should be present at the plus side of the voltmeter.
Answer D is wrong. If the voltmeter leads were to be reversed using a DMM, the display would simply show a negative value. If this were an analog type meter, the needle would attempt to move to the left of zero.

Question #7
Answer A is wrong because probing at the battery terminals would read battery voltage, not voltage drop in a circuit.
Answer B is wrong because this again is measuring battery voltage. The only difference is that the negative voltmeter probe is contacting ground, rather than the negative battery terminal.
Answer C is wrong because testing a circuit in series is only done when conducting current flow tests.
Answer D is correct because voltage drop tests are always performed by probing the leads in parallel with the component or circuit being tested.

Question #8
Answer A is wrong because ideally, all 12 volts from the battery should be "dropped" across the load regardless of the resistance of the bulb. A reading of 9 volts across the bulb indicates that 3 volts are being lost elsewhere in the circuit. This is the whole purpose of voltage drop testing.
Answer B is wrong because if there was a short to ground between the switch and the light, all the voltage would be going straight to ground (assuming the fuse did not blow) and there would be nothing left to be dropped across the light.
Answer C is wrong because neither technician is correct.
Answer D is correct because both technicians are wrong.

Question #9
Answer A is wrong. An ohmmeter is the best tool to use to test for shorts between circuits (with the circuit not under power). An ammeter measures current flow only; it cannot pinpoint a cross or short with another circuit.
Answer B is correct because low battery voltage will cause reduced current flow in a circuit with constant resistance (Ohm's law).
Answer C is wrong because only technician B is correct.
Answer D is wrong because one of the technicians is correct.

Question #10
Answer A is wrong because if the fuse were open, there would be no current flow at all through the circuit.
Answer B is wrong because reduced battery voltage will cause a reduction in current flow if the resistance of the load remains the same (Ohm's law).
Answer C is correct because a shorted bulb filament will cause reduced resistance in the bulb that will cause current flow to increase if the battery voltage remains constant (Ohm's law).
Answer D is wrong because an increase in bulb resistance will cause a reduction in current flow if the battery voltage remains constant (Ohm's law).

Question #11
Answer A is wrong because while an ammeter could theoretically be used to check continuity, it would be extremely impractical to do so. An ohmmeter is the best tool to use for this.
Answer B is correct because the sole purpose of an ammeter is to measure current flow.
Answer C is wrong because only technician B is correct.
Answer D is wrong because one of the technicians is correct.

Question #12
Answer A is wrong because you should never use an ohmmeter in a circuit where current is flowing. Erroneous readings and possible meter damage will result.
Answer B is wrong because in this case the scale should be set to ×1000, not ×100. Using the ×1000 scale will give much better resolution and accuracy.
Answer C is wrong because neither technician is correct.
Answer D is correct because both technicians are wrong.

Question #13
Answer A is correct because increased resistance in a circuit with constant battery voltage will cause reduced current flow (Ohm's law).
Answer B is wrong because a decrease in the resistance of a circuit will cause an increase in current flow with constant battery voltage applied (Ohm's law).
Answer C is wrong because an increase in battery voltage will cause an increase in current flow if the resistance of the load remains constant (Ohm's law).
Answer D is wrong because a short in the blower motor will cause reduced resistance, which will cause an increase in current flow with a constant battery voltage applied (Ohm's law).

Question #14
Answer A is wrong because you can have voltage present in a system without current flow. In that case, there can be no voltage drop.
Answer B is wrong because although resistance causes voltage drops, it need not be present in order to test a circuit for a voltage drop.
Answer C is correct because there must be current flow in a circuit to cause a voltage drop.
Answer D is wrong because tester probes must be placed in parallel to the circuit being tested, not in series.

Question #15
Answer A is a good choice because unwanted resistance will always produce heat in an electrical circuit; however, both technicians are right so answer A is wrong.
Answer B is also a good choice because increased resistance will always cause a decrease in current flow (Ohm's law). Since both technicians are right, answer B is also wrong.
Answer C is correct because both technicians are right.
Answer D is wrong because both technicians are right.

Question #16
Answer A is wrong because most DMMs have a 10–20 amp limit when making current tests directly through the meter.
Answer B is correct because most DMMs are not capable of measuring much more than 10–20 amps directly through a meter without blowing a fuse. For this reason an amp clamp should be used.
Answer C is wrong because only technician B is correct.
Answer D is wrong because one of the technicians is right.

Question #17
Answer A is wrong because a test lamp needs to have one end connected to battery or chassis ground in order to function properly.
Answer B is correct because a battery cannot be disconnected during a voltage test.
Answer C is wrong because the test light needs to have one end connected to either a chassis or battery ground in order to function properly.
Answer D is wrong because a weatherproof connector may be backprobed with the appropriate attachment that will allow a measurement to be taken without damaging the seal.

Question #18
Answer A is correct because if the test light remains on even after the connector is unplugged, this must mean that there is a short to ground somewhere between the test light and the unplugged connector. If there was no short to ground, the test lamp could not glow.
Answer B is wrong because if the circuit were open, there would be no way for current to flow through it and cause the test lamp to glow.
Answer C is wrong because only technician A is correct.
Answer D is wrong because one of the technicians is correct.

Question #19
Answer A is wrong because a self-powered test lamp should never be used in a circuit involving an electronic module. Most electronic circuits run very low voltages through them. Current from a higher voltage test light may damage it.
Answer B is wrong because an analog meter does not have the proper amount of impedance in it to allow the circuit to be accurately measured.
Answer C is wrong because neither technician is right.
Answer D is correct because both technicians are wrong.

Question #20
Answer A is wrong because the condition of the battery should always be checked first in this type of situation.
Answer B is wrong because the starter solenoid circuit would only be checked after making sure that the battery is properly charged.
Answer C is wrong because the ignition switch circuit should only be checked after verifying that the battery is fully charged.
Answer D is correct because checking for proper battery voltage is the first step in diagnosing this type of problem.

Question #21
Answer A is wrong because when the ignition switch is in the accessory position, there may be current being drawn through the battery to power certain loads that may be switched on. This would nullify a battery drain test.
Answer B is wrong because if the ignition switch is in the run position, there may be loads drawing current through the battery, causing an inaccurate battery drain test.
Answer C is wrong because if the ignition switch is in the crank position, current will be flowing through the battery, making a battery drain test meaningless.
Answer D is correct because with the ignition switch in the off position, all normal loads are disconnected from the battery circuit. This will allow for a proper battery drain test.

Question #22
Answer A is correct because a 2-amp draw through a battery with the engine off will cause the battery to discharge overnight.
Answer B is wrong because most electronic control modules draw less than 50 milliamps with the key off.
Answer C is wrong because only technician A is correct.
Answer D is wrong because one of the technicians is correct.

Question #23
Answer A is correct because a resistance has been added in series to the motor circuit. Ohm's law states that the current flow will decrease if the voltage stays the same.
Answer B is wrong. Because a resistance has been added to the circuit, current flow will decrease in this circuit, not increase.
Answer C is wrong. Ohm's law states that the current in the circuit will decrease, not stay the same.
Answer D is wrong because the circuit has not been broken (opened). Therefore there will be some current flow.

Question #24

Answer A is wrong. Never replace a fuse with one of a higher amperage rating. To do so may cause melted wiring and/or a fire.

Answer B is correct because a blown fuse usually indicates a short somewhere in the circuit causing an increase in current flow.

Answer C is wrong because if the circuit was open, current flow would stop and the fuse would not have blown in the first place.

Answer D is wrong because installing a circuit breaker of the same amperage rating as the fuse will not solve the original problem. The circuit breaker will merely cycle on and off.

Question #25

Answer A is correct. Disconnecting the battery first prevents accidental shorting of the wiring to ground while it is being repaired.

Answer B is wrong because a fuse link is always to be at least 2 wire gauge sizes smaller than the circuit it is protecting to allow for proper circuit protection.

Answer C is wrong because only technician A is correct.

Answer D is wrong because one of the technicians is correct.

Question #26

Answer A is wrong because a fusible link uses a special kind of insulation designed not to melt should there be an overload. Also, the fuse link should be at least 2 wire gauge sizes smaller than the circuit being protected for proper operation.

Answer B is wrong because circuit breakers are designed to be reset, either manually or automatically, after it trips.

Answer C is wrong because neither technician is correct.

Answer D is correct because both technicians are wrong.

Question #27

Answer A is wrong because a maxi fuse is simply a fuse much larger than standard.

Answer B is correct because maxi fuses are used in place of many fusible links on today's newer vehicles.

Answer C is wrong because a fuse is no good when it opens. A circuit breaker can be reset.

Answer D is wrong because not all truck electrical circuits are protected with maxi fuses. Many if not most truck circuits are protected with standard size fuses.

Question #28

Answer A is correct because this is a symbol for a diode.

Answer B is wrong because this is not a symbol for a resistor. Zigzag lines usually represent these.

Answer C is wrong because a capacitor is not pictured here.

Answer D is wrong because this is not a symbol for a thermistor.

Question #29

Answer A is correct. The arrow on a diode symbol points in the direction of normal current flow. The line next to the arrow point is to symbolize a blocked path to current flow. Thus, a diode allows current flow in only one direction.

Answer B is wrong because nowhere on the hot side of the bulb is there a short to ground. A diode is simply a one-way check valve.

Answer C is wrong because only technician A is correct.

Answer D is wrong because one of the technicians is correct.

Question #30
Answer A is wrong because the compressor clutch can only turn in one direction regardless of the application of the clutch. The driving engine cannot turn backwards.
Answer B is wrong because it is the A/C circuit that is being protected, not the clutch itself. The voltage spike is redirected back into the clutch during disengagement.
Answer C is correct because the purpose of the diode is to protect the rest of the circuit from a voltage spike generated when the clutch magnetic field collapses.
Answer D is wrong because the diode cannot limit current flow to the clutch because a diode is simply a one-way check valve, not a resistor. Even if the diode acted as a resistor, it would have to be wired in series with the clutch, not in parallel, to affect the current flow.

Question #31
Answer A is wrong. Control power in is pin #86.
Answer B is wrong. Control ground is pin #85.
Answer C is correct. Pin #30 is always used for high amperage power in.
Answer D is wrong. Normally closed power out is pin #87a.

Question #32
Answer A is wrong because relays are the perfect solution when a low current device (such as an ECM) needs to control a large current flow.
Answer B is wrong because solenoids are found in many different places other than a starter. One example would be a fuel shutoff solenoid.
Answer C is wrong because both technicians A and B are wrong.
Answer D is correct because both technicians are wrong.

Question #33
Answer A is wrong. The fuse is not open because the meter immediately after the fuse indicates 12 volts.
Answer B is correct because the voltmeter between the relay and ECM indicates 12 volts. If the ECM was functioning properly, the meter would read zero volts, because this would then be the ground side of the relay (load). Because the ECM is not providing the proper ground circuit, the meter reads battery voltage because there is no current flow.
Answer C is wrong. The relay cannot be condemned because the ECM is not allowing it to be activated.
Answer D is wrong because the fuel pump is not receiving any voltage from the relay as evidenced by the zero voltmeter reading next to the fuel pump. Therefore the pump cannot be considered faulty.

Question #34
Answer A is wrong because a battery's open circuit voltage must be at least 12.6 to be considered fully charged.
Answer B is wrong because this specification (13.5 to 14.5 volts) is for battery voltage while the engine is running, not for an open circuit battery test.
Answer C is wrong because neither technician is right.
Answer D is correct because both technicians are wrong.

Question #35
Answer A is wrong because the range is too low for a fully charged battery.
Answer B is wrong because this range is still below the threshold for a fully charged battery.
Answer C is correct because a fully charged battery would have a reading of at least 1.265.
Answer D is wrong because these values would indicate an overcharge or possible battery damage.

Question #36
Answer A is correct because a faulty voltage regulator may lead to an overcharging condition and possible battery boil over.
Answer B is wrong because a loose alternator belt might cause undercharging but not overcharging, which is what the low electrolyte level is pointing to here.
Answer C is wrong because only technician A is correct.
Answer D is wrong because one of the technicians is correct.

Question #37
Answer A is wrong because the proper discharge rate is 1/2 cold cranking amps or 3 times the ampere-hour rating.
Answer B is correct because a battery that remains above 9.6 volts after a 15-second test while still under load means that the battery has passed the test.
Answer C is wrong because only technician B is correct.
Answer D is wrong because one of the technicians is correct.

Question #38
Answer A is wrong because a battery parasitic drain test is performed with a voltmeter, not a carbon pile.
Answer B is correct because the figure shows a battery being prepared for a capacity test using a carbon pile to provide the load.
Answer C is wrong because a battery voltage test is done with a DMM, not a carbon pile.
Answer D is wrong because the state of charge test is done with a DMM or a hydrometer, not a carbon pile.

Question #39
Answer A is wrong because a battery that is being load tested must have a reading above 9.6 volts under load, not less, in order for it to pass the test.
Answer B is wrong because a battery should be load tested to 3 times its ampere-hour rating, not 2 times.
Answer C is wrong because neither technician is correct.
Answer D is correct because both technicians are wrong.

Question #40
Answer A is correct because a reading of 12.4 volts indicates a battery that is not fully charged. A reading of 12.6 volts would indicate a battery that is fully charged when tested in an open circuit state.
Answer B is wrong. Just because a battery is not fully charged does not mean that the battery needs replacement. This battery should be charged and then load tested to determine serviceability.
Answer C is wrong because only technician A is correct.
Answer D is wrong because one of the technicians is right.

Question #41
Answer A is wrong. Trucks with ECMs will constantly draw some power, even with the key off.
Answer B is correct because moisture on top of a battery can cause a surface discharge between the posts.
Answer C is wrong because only one of the technicians is correct.
Answer D is wrong because one of the technicians is correct.

Question #42
Answer A is correct. When disconnecting battery cables, always remove the ground cable first and reconnect it last to avoid sparks and a possible explosion.
Answer B is wrong because it is acceptable to clean battery tops and terminals with a baking soda and water solution.
Answer C is wrong because battery cable terminals should only be replaced with the proper crimp type terminals along with heat shrink tubing.
Answer D is wrong because it is good practice to coat terminal ends with a protective grease to retard corrosion.

Question #43
Answer A is correct because batteries can be damaged internally if they are not properly secured in the battery tray.
Answer B is wrong because batteries can self-discharge through any accumulated corrosion and moisture buildup across the top. A dirty battery tray will accelerate this process.
Answer C is wrong because only technician A is correct.
Answer D is wrong because one of the technicians is correct.

Question #44
Answer A is wrong because a fully charged battery will not readily accept a charge.
Answer B is wrong because a highly sulfated battery will not readily accept a charge due to high internal resistance.
Answer C is wrong because if there is a poor connection between the charging clamp and the battery terminal, excess resistance will prevent the battery from being charged.
Answer D is correct because even a heavy discharge across the top of the battery will still cause the ammeter on the charger to register a fair amount of current.

Question #45
Answer A is wrong because battery cables and their terminals can be reused if they are in good condition and pass a voltage drop test.
Answer B is correct because a charging system that is not operating properly can cause a battery failure. Always check for proper charging system operation when you suspect a battery failure.
Answer C is wrong because the starter motor should only be replaced if there is something wrong with it. It will have nothing to do with a battery failure.
Answer D is wrong because belt replacement will be determined during a charging system diagnosis.

Question #46
Answer A is wrong because a low battery most likely became low due to excessive cranking. This is when a battery generates most of its explosive vapors.
Answer B is correct because it is always considered good practice to wear eye protection when working near batteries.
Answer C is wrong because only technician B is correct.
Answer D is wrong because one of the technicians is correct.

Question #47
Answer A is wrong because you should never use compressed air to blow possible battery acid around.
Answer B is wrong because mineral spirits are not used to clean a battery.
Answer C is correct because you should always inspect and clean battery cable terminals if needed when servicing a battery to scrape away any accumulated corrosion.
Answer D is wrong because baking soda (a base) should be used to dissolve the residues on the batteries, not sulfuric acid.

Question #48
Answer A is wrong because the surface charge must be drawn off the battery if it has just come off the charger. Then the battery must be allowed to sit for 15 minutes before testing open-circuit voltage.
Answer B is wrong because a battery should never be tested for open-circuit voltage immediately after being charged or an erroneous reading will result.
Answer C is wrong because neither technician is correct.
Answer D is correct because both technicians are wrong.

Question #49
Answer A is wrong because the negative battery cable should always be disconnected first when charging a battery.
Answer B is wrong because you should consider a battery to be fully charged when the specific gravity reaches 1.265.
Answer C is wrong because you should reduce the fast charging rate when the specific gravity reaches 1.225.
Answer D is correct because if the battery is very low, it can freeze very easily. At this point, it should first be brought to room temperature before charging to prevent further damage.

Question #50
Answer A is wrong because a fast charge rate can overheat the battery if one is not very careful to monitor it during the process.
Answer B is correct because a slow charge rate will prevent possible battery damage due to overheating from a fast charge rate. This is also much safer because the battery will give off less gas during the process.
Answer C is wrong because a vehicle charging system will effectively fast charge a battery if it is very low. This is undesirable for the same reasons given in answer A.
Answer D is wrong. Electrolyte should never be added to a battery; only distilled water if it is low.

Question #51
Answer A is wrong because all unnecessary loads in both vehicles should be turned off to allow the maximum amount of current to flow to the battery and the starter.
Answer B is correct because the heavy spark generated when hooking up the last cable could ignite explosive vapors present around a heavily discharged battery. By hooking up the ground cable to an engine ground instead of the battery, this danger is eliminated.
Answer C is wrong because only technician B is correct.
Answer D is wrong because one of the technicians is right.

Question #52
Answer A is a good choice because by starting the booster vehicle, the charging system will produce more voltage to the battery than would be attainable if the engine were left off. This will aid in the starting process. However, since technician B is also correct, answer A is wrong.
Answer B is also a good choice because the engine should be off when connecting the cables to minimize the potential for a big spark and possible explosion. However, since technician A is also correct, answer B is wrong.
Answer C is correct because both technicians are correct.
Answer D is wrong because both technicians are right.

Question #53
Answer A is correct because a solenoid is typically used on top of a starter to engage the pinion to the flywheel and make the high current connection between the battery and starter motor.
Answer B is wrong because a ballast resistor is typically used only in ignition systems.
Answer C is wrong because a voltage regulator is part of the charging system.
Answer D is wrong because an ECM is normally not involved in starting an engine.

Question #54
Answer A is correct because a starter voltage drop test does not check the condition of a battery or its state of charge, only the resistance to current flow in the various connections and cables.
Answer B is wrong because resistance in the positive battery cables is one check that can be made during a starter circuit voltage drop test.
Answer C is wrong because resistance in the negative battery cables is one check that can be made during a starter circuit voltage drop test.
Answer D is wrong because resistance in the solenoid internal contacts is a check that can be made during a starter circuit voltage drop test.

Question #55
Answer A is wrong because a faulty magnetic switch would not provide power to the starter solenoid and make it click.
Answer B is wrong because a faulty key switch will not power the magnetic switch, which in turn will not power the starter solenoid and make it click.
Answer C is correct because faulty internal solenoid contacts will most likely prevent starter motor operation, however it would not prevent the solenoid plunger from moving into the engaged position and making the clicking noise.
Answer D is wrong because an open circuit between the starter solenoid and the magnetic switch would prevent solenoid operation altogether, therefore it would not click.

Question #56

Answer A is correct because when you hook up your voltmeter between the battery ground terminal and the base of the starter, you are effectively performing a voltage drop test of the ground side of the starting circuit. This is the same as checking the resistance of the ground side of this circuit.

Answer B is wrong because an ohmmeter cannot possibly circulate enough current through the starting circuit to determine excessive resistance. This can only be done with a voltage drop test.

Answer C is wrong because an ohmmeter should never be connected into a live circuit. Possible meter damage may result.

Answer D is wrong because in order to conduct a voltage drop test, the starter must be cranking. Also, testing the positive side of the battery will not allow the ground circuit to be checked.

Question #57

Answer A is wrong because while a faulty voltage regulator is the most likely cause, it can also be caused by a bad sense diode on systems with external regulators.

Answer B is wrong because excessive resistance in the charging circuit wiring should cause undercharging, not overcharging.

Answer C is wrong because neither technician is correct.

Answer D is correct because both technicians are wrong.

Question #58

Answer A is wrong because this would test the voltage drop in the positive battery cable.

Answer B is wrong because this would test the combined voltage drop of both the positive battery cable and the solenoid internal contacts.

Answer C is correct. By probing between points B and M, the voltage drop across the solenoid internal contacts can be tested with the starter cranking.

Answer D is wrong because probing between points G and ground will only test the voltage drop in the solenoid ground wire.

Question #59

Answer A is wrong because the magnetic switch would not click if it had a poor ground.

Answer B is correct. Even though the technician found 12 volts at the wire when he activated the key, the mistake he made was to check for voltage with the circuit open. Since there was no load, 12 volts was indicated even though the contacts were corroded in the magnetic switch. He should have probed for voltage at point S with the wire connected to have the circuit under a load.

Answer C is wrong because if the key switch contacts were faulty, the magnetic switch would have not clicked.

Answer D is wrong because if the battery ground cable had a poor connection, the magnetic switch would not have clicked.

Question #60

Answer A is wrong because the magnetic switch is only responsible for transmitting the current flow necessary to activate the starter solenoid.

Answer B is wrong because the key switch is only responsible for transmitting the current flow necessary to activate the magnetic switch.

Answer C is wrong because neither technician is right.

Answer D is correct because both technicians are wrong.

Question #61

Answer A is correct because the arrow is pointing to the start switch. The current from this switch flows to the solenoid pull-in and hold-in windings to activate the starter.

Answer B is wrong because this picture does not show a separate magnetic or relay switch.

Answer C is wrong because the pull-in winding is inside the solenoid, not external.

Answer D is wrong because the hold-in winding is inside the solenoid, not external.

Question #62
Answer A is wrong because the start switch is being bypassed in this test.
Answer B is wrong. Although the battery is being tested here, it is not the purpose of this test.
Answer C is correct. When terminals C and D are jumped, the magnetic switch is activated independent of the start switch. This causes voltage to be fed to the S terminal of the starter solenoid, which should cause it to engage.
Answer D is wrong. Although the starter is also being activated here, it is not the purpose of this test.

Question #63
Answer A is correct because this picture shows a magnetic switch.
Answer B is wrong because this figure does not represent a starting safety switch.
Answer C is wrong because this figure does not represent a starting switch.
Answer D is wrong because this figure does not represent a starter solenoid.

Question #64
Answer A is a good choice because the starting safety switch is used to prevent an engine from starting with the transmission in gear. However, technician B is also correct, so answer A is wrong.
Answer B is also a good choice because starting safety switches can also be called neutral safety switches. However, technician A is also correct, so answer B is wrong.
Answer C is correct because both technicians are correct.
Answer D is wrong because both technicians are right.

Question #65
Answer A is wrong because a starting switch is considered part of the control circuit.
Answer B is correct because the battery is not considered part of the control circuit. While it does supply the power for the circuit, it is not used to control the start circuit.
Answer C is wrong because the starting safety switch is definitely part of the control circuit. If it is not closed, the circuit is open.
Answer D is wrong because the magnetic switch is also part of the control circuit. Power from the control circuit is fed to the magnetic switch to activate the starter solenoid.

Question #66
Answer A is wrong because if the solenoid were not activated, the motor would not spin.
Answer B is correct because a faulty drive may not spin the engine even though the motor is turning.
Answer C is wrong because only technician B is correct.
Answer D is wrong because one of the technicians is correct.

Question #67
Answer A is wrong because if battery power was not connected to the starter motor, the starter would not spin.
Answer B is wrong because if the solenoid does not pull the pinion into mesh with the engine flywheel, the engine will not crank.
Answer C is wrong because a low current circuit is supposed to operate a large current circuit in order to start an engine. This is the function of the starter solenoid and the magnetic switch.
Answer D is correct because a drive pinion that fails to retract will not cause an engine to not start.

Question #68
Answer A is correct because in performing a voltage drop test on the insulated side of the starting circuit, one of the test leads needs to be probed on the positive battery cable while the starter is turning.
Answer B is wrong because the starter motor must be spinning to properly check for voltage drop.
Answer C is wrong because probing the negative battery cable checks the voltage drop on the ground side of the circuit, not the insulated side.
Answer D is wrong because it is not necessary to have the vehicle warmed up to perform a voltage drop test.

Question #69
Answer A is wrong because if the cranking motor turns but does not start, this usually indicates a problem with the drive pinion, not the brushes.
Answer B is wrong because if the pinion disengages slowly, this usually indicates a problem with the solenoid, not the brushes.
Answer C is wrong because noisy starter operation is usually associated with pinion problems, bent armatures, or bad bearings, not bad brushes.
Answer D is correct because bad brushes could cause high resistance in the circuit and, therefore, slow cranking speed.

Question #70
Answer A is wrong because a weak battery could cause an engine to crank slowly due to the reduced voltage.
Answer B is wrong because seized pistons or bearings could cause an engine to crank slowly because of the added resistance to turning the flywheel.
Answer C is correct because low resistance in the starter circuit will not cause an engine to crank slowly. Low resistance in the starter circuit is an ideal condition.
Answer D is wrong because high resistance in the starter circuit will cause excessive voltage drop and slower starter motor cranking speed.

Question #71
Answer A is wrong because a loose alternator belt can cause belt slippage, alternator undercharging, and, consequently, a discharged battery.
Answer B is wrong because a dirty battery cable connection can cause increased resistance at that point. The resulting voltage drop at this point could cause an undercharged battery because not all of the available alternator voltage can reach the battery to properly charge it.
Answer C is correct because a bad starter solenoid will have no direct affect on a discharged battery.
Answer D is wrong because a parasitic drain on the battery can definitely cause a battery to discharge over a period of time.

Question #72
Answer A is a good choice because in an alternator with an open field circuit, there will be no output due to no magnetic field inside the alternator. However, since technician B is also correct, answer A is wrong.
Answer B is also a good choice because if the fusible link between the battery and alternator was open, there would be no way for the alternator's output to reach the battery. However, since technician A is also correct, answer B is wrong.
Answer C is correct because both technicians are right.
Answer D is wrong because both technicians are correct.

Question #73
Answer A is correct because even though the belt is at the specified tension, it may slip because it is bottomed in the pulley.
Answer B is wrong because a misaligned pulley should not cause low alternator output because of belt slippage. It may, however, cause the belt to jump off, in which case there would be zero output.
Answer C is wrong because only technician A is correct.
Answer D is wrong because one of the technicians is correct.

Question #74
Answer A is a good choice because a loose drive belt can cause the pulley to slip with resultant undercharging. However, technician B is also correct, so answer A is wrong.
Answer B is also a good choice because undersized wiring between the battery and the alternator can result in excessive voltage drop that can result in an undercharged battery. However, technician A is also correct, so answer B is wrong.
Answer C is correct because both technicians are right.
Answer D is wrong because both technicians are correct.

Question #75
Answer A is wrong because the accessories should be off to allow the alternator to direct all of its output first to the battery and then to the carbon pile.
Answer B is wrong because the charging system voltage should be limited to 15.5 volts maximum to prevent damage to any electrical system components.
Answer C is wrong because neither technician is right.
Answer D is correct because both technicians are wrong.

Question #76
Answer A is wrong because this would be testing the voltage drop in the regulator field circuit.
Answer B is wrong because this would be testing the voltage drop on the ground side of the charging circuit, not the insulated side.
Answer C is correct because this is the correct test procedure to determine voltage drop on the insulated side of the charging circuit.
Answer D is wrong because this test would be measuring the voltage of the field circuit.

Question #77
Answer A is wrong because this test would be measuring the voltage drop in the regulator ground circuit.
Answer B is correct because this is the proper test procedure for measuring voltage drop on the ground side of the charging circuit.
Answer C is wrong because only technician B is right.
Answer D is wrong because one of the technicians is correct.

Question #78
Answer A is wrong because an ammeter cannot reliably detect very small amounts of battery discharge. Therefore the battery could go dead and the operator would be unaware of it.
Answer B is correct because a voltmeter is a more reliable indicator of system condition. If the meter reads 13.5 to 14.5 volts even under the heaviest of vehicle loads, then it is safe to say the charging system is operating properly.
Answer C is wrong because only technician B is correct.
Answer D is wrong because one of the technicians is correct.

Question #79
Answer A is wrong. A visual inspection will usually not reveal the cause of excessive resistance.
Answer B is correct. A voltage drop test is the best way to check a circuit for excessive resistance.
Answer C is wrong. An ohmmeter cannot accurately test resistance in large diameter conductors.
Answer D is wrong. Battery voltage would have to be compared simultaneously against alternator output for the results to be meaningful.

Question #80
Answer A is wrong because you should service battery cables as part of regular maintenance or whenever servicing the battery.
Answer B is wrong because battery corrosion can form on terminals in any weather. However, it affects the system most in cold weather.
Answer C is wrong because neither technician is correct.
Answer D is correct because both technicians are wrong.

Question #81
Answer A is wrong because there is no voltage solenoid in a truck charging system.
Answer B is correct because a voltage regulator is an integral part of the charging system, whether it is externally or internally mounted in the alternator.
Answer C is wrong because there is no voltage transducer in a truck charging system.
Answer D is wrong because a magnetic switch is part of the cranking circuit, not the charging circuit.

Question #82
Answer A is correct because a full-fielded alternator will cause maximum output from the alternator and consequent high battery voltage.
Answer B is wrong because a full-fielded alternator can cause excessive output voltage that could lead to burned out bulbs on the vehicle.
Answer C is wrong because maximum output from an alternator due to full-fielding can cause a battery to boil over due to excessive voltage.
Answer D is wrong. A full-fielded alternator will cause excessive battery voltage because the alternator is at maximum output.

Question #83
Answer A is correct because most headlight switches contain an internal circuit breaker that could trip if there is excessive current draw in the circuit. After it shuts down the circuit and cools, it automatically resets itself.
Answer B is wrong because you should never replace a circuit breaker or fuse with one of a larger capacity in order to solve a problem involving excessive current draw. To do so may cause wiring damage and/or a fire.
Answer C is wrong because only technician A is correct.
Answer D is wrong because one of the technicians is correct.

Question #84
Answer A is wrong because the headlight switch in normal operation does not normally control interior lights.
Answer B is correct because a poor ground at the door jam switch will cause excessive resistance in the circuit that will cause the light to glow dimly.
Answer C is wrong because only technician B is correct.
Answer D is wrong because one of the technicians is correct.

Question #85
Answer A is wrong because a faulty turn signal flasher will affect the trailer lights on both sides, not just one.
Answer B is wrong because it is not likely that all the bulbs on the left side have burned out. They need to be tested to be sure that they are getting proper voltage before being replaced.
Answer C is correct because the trailer electrical connector is one area where the two different circuits are independent of each other. If the pin for the left turn signal circuit has corrosion or is broken, this would be a good reason for the lights on the left hand side to be inoperative.
Answer D is wrong because a faulty brake light switch should affect both turn signal circuits in the same manner.

Question #86
Answer A is wrong because even if the voltage drop was higher than the desired maximum of 100 millivolts, this would reduce current flow in the circuit, not increase it. This reduced current flow should not cause a flasher failure, although increased current flow could.
Answer B is wrong because a poor ground at the trailer socket should cause reduced current flow, not increased current flow. For this reason, it should not cause a flasher unit to fail.
Answer C is wrong because neither technician is correct.
Answer D is correct because both technicians are wrong.

Question #87

Answer A is correct because temperature senders do not have voltage outputs like other senders do. They merely alter their resistance as the temperature changes and send that signal to an ECM or mechanical gauge.

Answer B is wrong because a good way to test a temperature sensor is to measure its resistance at a specific temperature and compare it against specifications.

Answer C is wrong because a good way to test a temperature sender is to immerse it in boiling water and see if the gauge reads 212°F (100°C).

Answer D is wrong because another good way to test a temperature sender for accuracy is to compare its reading against other temperature senders in the truck after it has been sitting overnight. If they all read within a few degrees of each other, it is safe to say they are all accurate.

Answers to the Test Questions for the Additional Test Questions Section 6

1.	D	28.	A	55.	A	82.	C
2.	D	29.	D	56.	B	83.	B
3.	C	30.	D	57.	C	84.	C
4.	A	31.	C	58.	D	85.	C
5.	A	32.	D	59.	A	86.	A
6.	C	33.	C	60.	B	87.	D
7.	B	34.	D	61.	B	88.	A
8.	A	35.	D	62.	C	89.	D
9.	B	36.	C	63.	D	90.	B
10.	A	37.	C	64.	C	91.	D
11.	B	38.	B	65.	C	92.	A
12.	D	39.	A	66.	B	93.	B
13.	A	40.	D	67.	A	94.	A
14.	D	41.	C	68.	C	95.	B
15.	D	42.	B	69.	C	96.	D
16.	C	43.	B	70.	D	97.	C
17.	C	44.	A	71.	C	98.	B
18.	D	45.	B	72.	A	99.	D
19.	C	46.	D	73.	C	100.	A
20.	A	47.	A	74.	D	101.	C
21.	C	48.	B	75.	D	102.	D
22.	D	49.	C	76.	A	103.	D
23.	B	50.	B	77.	A	104.	B
24.	D	51.	A	78.	C	105.	B
25.	A	52.	D	79.	A	106.	B
26.	B	53.	D	80.	B	107.	A
27.	C	54.	B	81.	A		

Explanations to the Answers for the Additional Test Questions Section 6

Question #1
Answer A is wrong because a starter drive pinion is supposed to have a chamfer on the drive teeth to assist in meshing with the flywheel ring gear.
Answer B is wrong because a flywheel ring gear can be replaced without replacing the entire flywheel.
Answer C is wrong because neither technician is correct.
Answer D is correct because both technicians are wrong.

Question #2
Answer A is wrong because overcharging will cause a battery to boil out electrolyte, and will also damage internal plates if it is severe enough.
Answer B is wrong because a battery will accept a charge much easier in warmer weather, requiring less voltage to do so.
Answer C is wrong because both technicians A and B are wrong.
Answer D is correct because neither technician is correct.

Question #3
Answer A is wrong because in order for the voltmeter to read charging system voltage, the negative lead of the voltmeter would need to be probing the battery negative terminal.
Answer B is wrong because the regulator in this example has been bypassed.
Answer C is correct because the voltmeter has been connected across either end of the charging system output circuit, consequently the voltage drop in this circuit can be measured with the system operating at rated output.
Answer D is wrong because to measure the voltage drop across the ignition switch, the voltmeter would need to be probed between a power in and a power out terminal on the switch itself.

Question #4
Answer A is correct. Disconnecting the positive battery cable first will generate sparks and possibly a battery explosion if the wrench contacts ground.
Answer B is wrong. Always reconnect battery cables in the reverse order of removal to avoid sparks and a possible explosion.
Answer C is wrong because only technician A is correct.
Answer D is wrong because one of the technicians is correct.

Question #5
Answer A is correct because most systems operate by turning on the charge warning light with the key on, engine off.
Answer B is wrong. Just because the indicator light is off with the engine running, the alternator might not be keeping up with all of the truck's load demands if the output is not up to specification.
Answer C is wrong because only technician A is correct.
Answer D is wrong because one of the technicians is correct.

Question #6
Answer A is a good choice because a blown fuse will interrupt the current flow to the bulb, causing it to not come on. However, technician B is also correct, so answer A is wrong.
Answer B is also a good choice because a burned-out bulb would obviously prevent the bulb from lighting in the event of a charging system malfunction. However, technician A is also correct, so answer B is wrong.
Answer C is correct because both technicians are right.
Answer D is wrong because both technicians are correct.

Question #7
Answer A is wrong because the dash voltmeter reads battery voltage.
Answer B is correct because resistance in the voltmeter circuit will cause a voltage drop at that point, which will affect the accuracy of the gauge.
Answer C is wrong because only one of the technicians is correct.
Answer D is wrong because one of the technicians is correct.

Question#8
Answer A is correct because the picture shows a diode being checked for continuity with an ohm-meter. In practice, the leads should then be reversed. It should show continuity in one direction, but not the other.
Answer B is wrong because in order to test voltage drop across a component, there must be current flowing in a complete circuit. Also, a voltmeter would be used for this test, not an ohmmeter.
Answer C is wrong because only technician A is correct.
Answer D is wrong because one of the technicians is correct.

Question #9
Answer A is wrong because replacing the entire battery cable is an acceptable method of repair. It may be the fastest and most economical way depending on shop preferences.
Answer B is correct because you should never replace a battery cable end with an aftermarket type bolt on end. These have poor wire contact characteristics and the entire joint is exposed and subject to corrosion.
Answer C is wrong because replacing a battery cable terminal end with a crimp on terminal and heat shrink tubing is considered an acceptable method of repair.
Answer D is wrong because replacing a battery cable terminal end with a soldered on terminal and heat shrink tubing is considered an acceptable method of repair.

Question #10
Answer A is correct because a carbon pile tester can safely and quickly load the system in excess of the alternator's maximum output. This will force the alternator to go to maximum output.
Answer B is wrong. Even though full-fielding (where applicable) is an easy way to make the alternator put out maximum output, it is a potentially dangerous test in that if the current is not dissipated somewhere, voltage in the system can rise to dangerous levels.
Answer C is wrong because all of the vehicle loads combined should not be equal to or exceed the alternator's maximum output if the system was properly engineered.
Answer D is wrong because temporarily installing a low battery will not necessarily force the alternator to maximum output. This would depend on the condition of the battery and how low it is.

Question #11
Answer A is wrong because voltmeters are always connected in parallel with the circuit being tested, not in series.
Answer B is correct because a carbon pile can draw current from the battery in excess of the alternator's output, whereas vehicle loads cannot. This will force the alternator to put out maximum output.
Answer C is wrong because only technician B is correct.
Answer D is wrong because one of the technicians is correct.

Question #12
Answer A is wrong because an ammeter is also needed to determine the current output of the alternator.
Answer B is wrong because a carbon pile is needed to load the system in excess of the alternator's output to force it to maximum output.
Answer C is wrong because both technicians are wrong.
Answer D is correct because neither technician is correct.

Question #13

Answer A is correct because the technician is performing a voltage drop test on the insulated (positive) side of the charging circuit. A 2-volt drop is far in excess of specifications, and consequently this will cause an undercharged battery.

Answer B is wrong because an excessive voltage drop will cause low voltage and dim lights, not high voltage and headlight flare-up.

Answer C is wrong because only technician A is correct.

Answer D is wrong because one of the technicians is correct.

Question #14

Answer A is wrong because a visual inspection will not always reveal internal problems with wiring or connections.

Answer B is wrong because an ohmmeter cannot circulate enough current through a large diameter wire to effectively tell you if it has excessive resistance.

Answer C is wrong because voltage drop tests are only meaningful with the circuit under normal full loads.

Answer D is correct because a voltage drop test under load is the best way to tell if a circuit has excessive resistance.

Question #15

Answer A is wrong because the negative battery cable should always be removed first to prevent accidental shorting of your wrench to ground.

Answer B is wrong because not all alternators have polarized connections, especially if they have been modified in the past. Be sure to label all multiple connectors before removing.

Answer C is wrong because neither technician is right.

Answer D is correct because both technicians are wrong.

Question #16

Answer A is wrong because this action is normal. The battery is being replenished quickly, and so the charging current going into it will steadily decrease as time goes on.

Answer B is wrong because a slipping drive belt should produce a consistent current output from the alternator.

Answer C is correct because this is a normal occurrence as a battery is recharged.

Answer D is wrong because diodes do not fail gradually. They either work or they do not.

Question #17

Answer A is wrong. A full-fielded alternator will produce maximum output regardless of voltage. This will cause the battery voltage to rise above safe levels.

Answer B is wrong. A battery that is being overcharged can cause the battery to boil and gas.

Answer C is correct because a full-fielded alternator will cause battery voltage to rise unregulated.

Answer D is wrong because a full-fielded alternator will cause very high amperage output from the alternator.

Question #18

Answer A is wrong because a bad ground in the left headlight would affect both high and low beam operation.

Answer B is wrong because high resistance in the dimmer switch high beam contacts would affect both right- and left-side bulbs.

Answer C is wrong because neither technician is correct.

Answer D is correct because both technicians are wrong.

Question #19
Answer A is a good choice because headlight aim should be checked on a level floor with the vehicle unloaded, but technicians A and B are both correct so answer A is wrong.
Answer B is also a good choice because this may conflict with existing laws and regulations in some states, but technicians A and B are both correct so answer B is wrong.
Answer C is correct because both technicians are right.
Answer D is wrong because both technicians are correct.

Question #20
Answer A is correct because when headlight aiming equipment is not available, headlight aiming can be checked by projecting the upper beam of each light upon a screen or chart at a distance of 25 feet ahead of the headlights.
Answer B is wrong because the vehicle should be exactly perpendicular to the chart or screen.
Answer C is wrong because only technician A is correct.
Answer D is wrong because technician A is correct.

Question #21
Answer A is wrong because corrosion on the connector will create a voltage drop which will cause a dim headlight.
Answer B is wrong because a damaged headlight assembly can cause increased resistance at the connections or bulb filament, causing a voltage drop and dim lights.
Answer C is correct because low alternator output will cause both lights to be dim.
Answer D is wrong because high resistance in the wiring leading to the bulb will cause a voltage drop and therefore decreased light output.

Question #22
Answer A is wrong because an open should always be checked upstream (toward the source) from an inoperative component.
Answer B is wrong because an ammeter is used to check current flow. It is very impractical to use it as a method of checking continuity.
Answer C is wrong because both technicians A and B are wrong.
Answer D is correct because both technicians A and B are wrong.

Question #23
Answer A is wrong because the taillight circuit uses only one pin in the trailer connector. Therefore, a faulty pin would have to affect the taillights on both sides equally.
Answer B is correct because a poor ground connection at the left-rear taillight may cause excessive resistance and low-current flow, therefore causing a dim light.
Answer C is wrong because only technician B is correct.
Answer D is wrong because one of the technicians is correct.

Question #24
Answer A is wrong because the park position on this switch would be the #2 position.
Answer B is wrong because there is no "run" position for a headlight switch.
Answer C is wrong because the on position for this switch would be the #3 position.
Answer D is correct because position #1 on this switch represents the "off" position.

Question #25
Answer A is correct because dielectric lubricant is the recommended lube for almost all electrical connections.
Answer B is wrong because lithium grease is not recommended for electrical connections.
Answer C is wrong because technician B is wrong.
Answer D is wrong because technician A is correct.

Question #26
Answer A is wrong because a voltage limiter is used to regulate power to instrument gauges at a constant voltage level. There is no way to adjust the voltage output with one of these.
Answer B is correct because a rheostat is a variable resistor. By turning the rheostat, the resistance is either increased or decreased, thereby changing the voltage and current flow to the bulb and altering its brightness.
Answer C is wrong because a resistor has a fixed resistance value and cannot be changed.
Answer D is wrong because a diode is simply a one-way electrical check valve that allows current flow in one direction but not the other. It will not affect the resistance of the circuit in the direction of current flow.

Question #27
Answer A is a good choice because an alternator that is overcharging will cause excessive system voltage that will increase the current flow through the lights causing them to be brighter than normal. However, technician B is also correct, so answer A is wrong.
Answer B is also a good choice because poor chassis grounds are usually the cause of dim lights due to rust and corrosion at mounting points. However, technician A is also correct, so answer B is wrong.
Answer C is correct because both technicians are correct.
Answer D is wrong because both technicians are right.

Question #28
Answer A is correct because headlight dimmer switches can be floor mounted.
Answer B is wrong because headlight dimmer switches are not mounted on the front panel. They need to be in a place quickly found by the driver in the dark.
Answer C is wrong because only technician A is right.
Answer D is wrong because one of the technicians is correct.

Question #29
Answer A is wrong because a faulty dimmer switch can cause faulty high beam operation.
Answer B is wrong because a loose or otherwise faulty high beam filament will cause partial or total failure of the high beam circuit.
Answer C is wrong because a loose connector in the high beam circuit can affect its operation.
Answer D is correct because a faulty headlight switch should affect both low and high beam operation equally, since it supplies power for both circuits out of the same set of contacts.

Question #30
Answer A is wrong because if the short was upstream of the bulb holder, the fuse would blow regardless of whether the bulb was in place or not.
Answer B is wrong because a short to ground on the ground side of the circuit will not affect the operation of this bulb.
Answer C is wrong because technicians A and B are both wrong.
Answer D is correct because neither technician is correct.

Question #31
Answer A is a good choice because multifunction switches may incorporate cruise controls into them. However, technician B is also correct, so answer A is wrong.
Answer B is also a good choice because a multifunction switch also serves as a turn signal switch. However, technician A is also correct, so answer B is wrong.
Answer C is correct because both technicians are correct.
Answer D is wrong because both technicians are right.

Question #32
Answer A is wrong because an open in the trailer connector would cause both taillights to be inoperative.
Answer B is wrong because an ohmmeter cannot be used to check continuity of a circuit under power.
Answer C is wrong because both technicians are wrong.
Answer D is correct because neither technician is right.

Question #33
Answer A is a good choice because halogen headlight bulbs will not tolerate oil on them from your skin. However, technician B is also correct, so answer A is wrong.
Answer B is also a good choice because halogen bulbs will outlast conventional sealed beams. However, technician A is also correct, so answer B is wrong.
Answer C is correct because both technicians are right.
Answer D is wrong because both technicians are correct.

Question #34
Answer A is wrong because the courtesy light would not work at all with a blown cigar lighter fuse.
Answer B is wrong because the stop and dome lights receive their power from a separate fuse.
Answer C is wrong because the parking, instrument, and taillights receive their power from a separate fuse.
Answer D is correct because the courtesy light sources its power from the cigar lighter fuse.

Question #35
Answer A is wrong because the parking lights receive their power from the headlight switch.
Answer B is wrong because the stoplights do not source their power from the headlight switch.
Answer C is wrong because neither technician is correct.
Answer D is correct because both technicians are wrong.

Question #36
Answer A is wrong because circuit DB 180G RD feeds the left-hand rear wiring, not the right-hand. This can be determined by following the wiring to the turn signal switch, where it terminates at a point marked "LT," or left turn.
Answer B is wrong because a short to ground in this wire would affect the left-hand lights, not the right-hand side.
Answer C is correct because circuit D7 18BR RD feeds the right-hand rear wiring. This can be determined by following the wiring to the turn signal switch, where it terminates indirectly to a point marked "RT," or right turn. High resistance in this circuit would cause a dim right rear signal light.
Answer D is wrong because high resistance in the wiring from the turn signal flasher unit to the turn signal switch would affect both left and right lamps the same.

Question #37
Answer A is a good choice because if the front bulb fails, this will reduce the current demands in the circuit, causing the flasher to blink slower than normal. Yet, this is wrong because both technicians are right.
Answer B is also a good choice because an open circuit going to the front bulb will also reduce the current demands in the circuit, making the flasher blink slower than normal and also not allow the front bulb to light. Yet, this is wrong because both technicians are right.
Answer C is correct because both technicians are right.
Answer D is wrong because both technicians are correct.

Question #38
Answer A is wrong because if the fuse keeps blowing, this means that there must be a short somewhere to ground, or a faulty bulb. Checking between points A and B will not indicate a short to ground.
Answer B is correct because if a short to ground is suspected, then the ohmmeter must be placed at some point in the circuit with respect to ground. An unusually low reading will indicate a short to ground problem.
Answer C is wrong because only one of the technicians is correct.
Answer D is wrong because one technician is correct.

Question #39
Answer A is correct because a short to ground on the hot side of the bulb will cause increased current flow that will blow the fuse.
Answer B is wrong because even if the fuse did not blow, the increased current demand in the circuit on the right-hand side would cause the left-hand bulb to be dim. This would happen because the two branches would no longer have equal resistance; rather, the right-hand side would have less resistance and consequent increased current flow.
Answer C is wrong because only one of the technicians is correct.
Answer D is wrong because one of the technicians is correct.

Question #40
Answer A is wrong because a stoplight switch is usually located on the brake pedal on medium-duty trucks with hydraulic brakes.
Answer B is wrong because the current from a stoplight switch is most often routed through the turn signal switch to allow for proper operation of the brake lights and turn signals simultaneously.
Answer C is wrong because neither technician is correct.
Answer D is correct because both technicians are wrong.

Question #41
Answer A is a good choice, because the same fuse that powers the taillights most likely also provides power for the dash lights. However, both technicians are correct, so answer A is wrong.
Answer B is also a good choice because a faulty rheostat could cause the dash lights not to come on. However, both technicians are correct, so answer B is wrong.
Answer C is correct because both technicians are correct.
Answer D is wrong because both technicians are correct.

Question #42
Answer A is wrong because a short to ground on the ground side of the bulb will not change overall circuit resistance. Therefore, current flow will not increase.
Answer B is correct because a short to ground on the ground side of this bulb has effectively made the switch useless. This particular circuit uses ground side switching, but in this case, the short has bypassed the switch.
Answer C is wrong because only technician B is correct.
Answer D is wrong because one of the technicians is right.

Question #43
Answer A is wrong because older bi-metal type gauges are sensitive to voltage fluctuations. For this reason, they use what is known as an IVR (instrument voltage regulator) to maintain constant voltage to the gauges regardless of battery voltage to keep gauge accuracy constant.
Answer B is correct because most magnetic gauges do not require a separate IVR to maintain constant voltage.
Answer C is wrong because only technician B is correct.
Answer D is wrong because one of the technicians is correct.

Question #44
Answer A is correct because high resistance in the sender ground circuit will cause reduced current flow, which would cause the heating coil to not bend the bimetallic arm as far as it should.
Answer B is wrong because a high resistance after the voltage regulator should affect all the other gauges equally.
Answer C is wrong because a short to ground between the gauge and sender will cause reduced resistance and increased current flow, causing the heating coil to get much hotter, therefore bending the bimetallic arm much further than normal.
Answer D is wrong because an open circuit between the gauge and sending unit will cause zero current flow and no needle movement.

Question #45
Answer A is wrong because the instrument voltage limiter maintains a constant voltage supply to the gauges regardless of battery voltage.
Answer B is correct because a faulty instrument voltage limiter will affect all gauges in the same manner.
Answer C is wrong because only technician B is correct.
Answer D is wrong because one of the technicians is correct.

Question #46
Answer A is wrong because not all gauges are affected by the voltage limiter. Some gauges are mechanical, such as the air application gauge.
Answer B is wrong because other gauges in the panel that share the same voltage source would be similarly affected.
Answer C is wrong because not all gauges are affected by the voltage limiter. Some gauges are mechanical, such as the air application gauge.
Answer D is correct because a malfunctioning voltage limiter would only affect those gauges to which it sends power.

Question #47
Answer A is correct because the ground for the fuel level gauge is independent of any other grounds in the instrumentation circuit. Therefore only the fuel gauge will be affected.
Answer B is wrong because although all the gauges share the same ground circuit, it is not the same ground circuit as the one for the fuel level sender.
Answer C is wrong because only technician A is correct.
Answer D is wrong because only one of the technicians is wrong.

Question #48
Answer A is wrong because the electronic instrument panel can only source its information from the data bus via the main ECM.
Answer B is wrong because the electronic instrument panel can also source its information from processors other than the main ECM.
Answer C is wrong because neither technician is correct.
Answer D is correct because neither technician is right.

Question #49
Answer A is wrong. While swapping panels may help to locate a faulty gauge, it is false economy because electrical/electronic components are not returnable if not needed.
Answer B is wrong. As with the previous answer, it is bad practice to "chase" a problem with new components until you locate the problem.
Answer C is wrong because with many senders on an electronic engine, grounding is not possible.
Answer D is correct. By using the diagnostic tool, you can compare what the ECM is reporting over the data bus with what the actual reading is at the gauge in question.

Question #50
Answer A is wrong because the ignition switch only provides basic power to the circuits; it is not responsible for switching individual lights on and off.
Answer B is correct because closing a switch or sensor is usually required to complete a circuit to power a light or warning device.
Answer C is wrong because opening a switch or sensor will interrupt current flow in a circuit, thereby canceling the operation of a light or warning device.
Answer D is wrong because the vehicle battery only supplies power to operate a particular device; it does not and cannot turn a particular device on and off.

Question #51
Answer A is wrong because if the rear axle ratio has not been properly programmed into the engine computer, the speedometer readings will be inaccurate.
Answer B is correct. The transmission speed sensor does not need to be calibrated; it merely sends a report back to the engine computer of the number of output shaft revolutions it sees. It is necessary for the other operating parameters to be properly entered in order to get an accurate speedometer reading.
Answer C is wrong because if the tire rolling radius has not been properly entered into the engine computer, an erroneous speedometer reading will result.
Answer D is wrong because some engine management systems keep track of tire wear. If new tires are installed and the engine computer is not "informed" of this, false speedometer readings will result.

Question #52
Answer A is wrong because while many gauges can be checked this way, doing so might damage some gauges. Always consult with the vehicle repair manual before using this procedure.
Answer B is correct because a variable resistance test box is a good way to substitute a resistance value to the gauge that would otherwise have to come from a questionable sender unit. By comparing against the manufacturer's specifications, the appropriate value can be substituted for any expected gauge reading.
Answer C is wrong because only technician B is correct.
Answer D is wrong because only one of the technicians is wrong.

Question #53
Answer A is wrong because pyrometers do not measure fuel flow.
Answer B is wrong because a voltmeter measures the battery state of charge, not a pyrometer.
Answer C is wrong because a tachometer displays engine speed, not a pyrometer.
Answer D is correct because it is the purpose of a pyrometer to measure temperatures.

Question #54
Answer A is wrong because checking gauge resistance is not considered an accurate way to test gauge accuracy.
Answer B is correct because this is a very accurate way to test a temperature gauge.
Answer C is wrong because only technician B is correct.
Answer D is wrong because one of the technicians is right.

Question #55
Answer A is wrong because low engine oil pressure would require an immediate action from the driver.
Answer B is wrong because high engine water temperature would also require an immediate response from the driver.
Answer C is correct. A maintenance reminder is a good example of a situation that did not require the driver's immediate attention.
Answer D is wrong because low coolant level is a potentially damaging engine condition that requires immediate attention.

Question #56
Answer A is wrong because the right side cluster of gauges does not source its information from the engine ECM.
Answer B is wrong because the right side cluster of gauges does not source its information from the engine ECM, although the middle cluster does.
Answer C is wrong because the middle cluster of gauges and most of the warning lamps gets its information from the engine ECM via the data bus.
Answer D is correct because most of the gauges and warning lights in those clusters receive their information from the engine ECM. The information for the gauges in the right hand cluster is likely sourced from another vehicle or chassis ECM.

Question #57
Answer A is wrong because this gauge action indicates that the dash has performed a self-test to check the operation of all the gauges.
Answer B is wrong because it is not very likely that battery voltage would be too high when the key is first turned on.
Answer C is wrong because neither technician is correct.
Answer D is correct because both technicians are wrong.

Question #58
Answer A is wrong because the gauge should only read high if the sender wire was shorted to ground, not open.
Answer B is wrong because if circuit 924 was open, the gauge should not read at all.
Answer C is wrong because an open in circuit 924 should cause no action at all, not fluctuations.
Answer D is correct because an open in the sender circuit should cause the gauge to be inoperative because there can be no current flow through the circuit.

Question #59
Answer A is a good choice because vehicle gauge instrumentation is generally accurate enough for most troubleshooting procedures. However, technician B is also correct, so answer A is wrong.
Answer B is also a good choice because you should never base major engine repair decisions on potentially false gauge readings without first verifying the information. However, technician A is also correct, so answer B is wrong.
Answer C is correct because both technicians are correct.
Answer D is wrong because both technicians are right.

Question #60
Answer A is wrong because a faulty ground at the sender unit could alter the circuit's overall resistance and throw off the gauge reading.
Answer B is correct because high battery voltage is compensated for by the IVR (instrument voltage regulator). Most magnetic gauges are not affected by varying voltage levels.
Answer C is wrong because a faulty instrument voltage regulator could allow too much voltage to bi-metal gauges, resulting in an inaccurate reading.
Answer D is wrong because excessive resistance in the wiring will alter the circuit's overall resistance, throwing off the gauge readings.

Question #61
Answer A is wrong because test lights draw too much current from an electronic circuit. The resulting reduction in resistance and increased current draw in that circuit will alter the circuit's measurable values.
Answer B is correct. Test lights are very useful tools when diagnosing complaints with accessories such as headlights and horns, where the current draw of the test light would only be a fraction of the whole circuit.
Answer C is wrong because only technician B is correct.
Answer D is wrong because one of the technicians is correct.

Question #62
Answer A is wrong because tachometers can derive their signal from the main data bus.
Answer B is wrong because tachometers can source their signal from the back of an alternator phase tap.
Answer C is correct because tachometers do not get speed signals from injector driver units.
Answer D is wrong because a tachometer can get a signal generated from an engine-driven sensor.

Question #63
Answer A is correct because the magnetic pickup shown in the picture is used for speed sensing, such as tachometers and speedometers.
Answer B is wrong because the picture does not show a pressure sensor.
Answer C is wrong because the figure does not show a temperature sensor.
Answer D is wrong because the figure does not show a level sensor.

Question #64
Answer A is wrong because 2 psi is within the acceptable tolerance level for testing air gauges.
Answer B is wrong because 8 psi is not within the allowable range for testing air gauges.
Answer C is correct because 4 psi is the maximum allowable error when testing gauge units.
Answer D is wrong because 10 psi is outside the allowable range for testing air gauges.

Question #65
Answer A is wrong because a diesel engine does not use a primary ignition coil.
Answer B is wrong because a diesel engine does not use a secondary ignition coil.
Answer C is correct because some tachometers can derive their signal off the back of the alternator.
Answer D is wrong because a vehicle speed sensor provides a signal to the speedometer, not the tachometer.

Question #66
Answer A is wrong because welded diaphragm contacts inside the horn will not allow it to vibrate and make sound.
Answer B is correct because a faulty relay can cause a horn to sound constantly if it is stuck in the energized position.
Answer C is wrong because technician B is right.
Answer D is wrong because one of the technicians is right.

Question #67
Answer A is correct. An open ground will not cause this condition because it is the action of the horn button to ground the relay, so no ground is normal.
Answer B is wrong because an open circuit in the horn relay winding will cause no horn operation.
Answer C is wrong because an open circuit at the horn brush/slip ring will cause no horn operation.
Answer D is wrong because an open fusible link in the relay power wire will cause no horn operation.

Question #68
Answer A is correct. A short to ground at connector C206 would simulate the horn switch being depressed. This action completes the ground circuit for the coil in the horn relay, which energizes the relay. This sends current to the horn continuously.
Answer B is wrong. A short to ground at connector C100 would likely burn out one of the fusible links, but only when the horn switch was depressed, because this circuit is only energized at that time.
Answer C is wrong because only technician A is correct.
Answer D is wrong because only one of the technicians is wrong.

Question #69
Answer A is a good choice because a test lamp is a good way to test for proper power and grounds at the relay. However, technician B is also correct, so answer A is wrong.
Answer B is also a good choice because a DMM is another good way to test for power and grounds at the relay. However, technician A is also correct, so answer B is wrong.
Answer C is correct because both technicians are right.
Answer D is wrong because both technicians are correct.

Question #70
Answer A is wrong because a blown fuse would make the horn totally inoperative, not intermittent.
Answer B is wrong because no power to the relay would make the horn totally inoperative, not intermittent.
Answer C is wrong because an open in the horn button circuit makes the horn totally inoperative, not intermittent.
Answer D is correct because a faulty horn relay could cause intermittent horn operation if its action is erratic.

Question #71
Answer A is wrong because a faulty wiper switch would prevent current flow to the wiper motor and would not allow motor operation.
Answer B is wrong because a tripped thermal overload protector would open the circuit and prevent current flow to the motor.
Answer C is wrong because a tripped circuit breaker would open the circuit and not allow current flow to the motor.
Answer D is correct because resistance in the wiring should only slow the motor down, not stop it.

Question #72
Answer A is a good choice because some 2-speed wiper motors use high and low speed brushes. However, technician B is also correct, so answer A is wrong.
Answer B is also a good choice because some 2-speed wiper motors use an external resistor pack to control the speeds. However, technician A is also correct, so answer B is wrong.
Answer C is correct because both technicians are correct.
Answer D is wrong because neither technician is wrong.

Question #73
Answer A is a good choice because wiper systems can either be air or electric operated. However, technician B is also correct, so answer A is wrong.
Answer B is also a good choice because a wiper system can use one or two motors. However, technician A is also correct, so answer B is wrong.
Answer C is correct because both technicians are correct.
Answer D is wrong because neither technician is wrong.

Question #74
Answer A is correct because poor brush contacts inside the motor will increase circuit resistance. This will lower current flow and cause sluggish operation.
Answer B is wrong because an open in the motor ground circuit will stop all current flow. In this case, the motor would not operate at all.
Answer C is wrong because only technician A is correct.
Answer D is wrong because one of the technicians is correct.

Question #75
Answer A is correct. The function of the park switch is to stop the motor at the same place every time regardless of the position of the wiper switch.
Answer B is wrong because the thermal overload is what shuts down the wiper motor in case of overheating.
Answer C is wrong because it is the function of the linkage to keep the blades synchronized (single motor systems).
Answer D is wrong because there is no protection system for a wiper motor in a low voltage situation.

Question #76
Answer A is correct because the purpose of the compressor clutch diode is to act as a spike suppression device to protect sensitive electronic components from becoming damaged due to voltage spikes.
Answer B is wrong because the direction of compressor rotation is determined by the direction the pulley is rotating.
Answer C is wrong because the clutch diode is installed parallel with the clutch coil. Even if the diode failed, the compressor clutch will still receive voltage.
Answer D is wrong because the presence (or lack thereof) of a functioning diode will have no measurable effect on the circuit's current requirements.

Question #77
Answer A is correct because if the activation arm for the park switch is broken or out of adjustment, the wipers may not park.
Answer B is wrong because a faulty wiper switch only controls the on-off functions of the motor, not the park position.
Answer C is wrong because only technician A is correct.
Answer D is wrong because one of the technicians is correct.

Question #78
Answer A is correct because binding wiper linkage can cause no wiper operation.
Answer B is wrong because a shorted control circuit should blow the fuse, not cause it to operate constantly.
Answer C is wrong because only technician A is right.
Answer D is wrong because one of the technicians is right.

Question #79
Answer A is wrong. While theoretically the resistance of the wire could be measured with an ohmmeter, a voltage drop test is a far more accurate way of assessing the condition of this wire.
Answer B is wrong. An ammeter will only show the amount of current flowing through the circuit; it will not give a good indication of the resistance of the particular wire. Excessive resistance could be in the motor itself, but the ammeter would not help to isolate this.
Answer C is correct. A voltage drop test with the circuit under load is the best way to determine if excessive resistance is present.
Answer D is wrong. Simply measuring the voltage at the end of the wire would not isolate the problem to the wire itself. Low voltage could be due to a weak battery, but you could not detect this unless you measured voltage at both ends of the wire simultaneously. This is in effect what you're doing in a voltage drop test.

Question #80
Answer A is wrong because a blown fuse will also cause the wiper motor to be inoperative.
Answer B is correct because if the isolation diode has an open circuit, the washer pump circuit will also be open, hence no washer pump operation.
Answer C is wrong because only technician B is correct.
Answer D is wrong because one of the technicians is correct.

Question #81
Answer A is wrong because if the ground side of the motor shorted to ground, there would be no adverse reaction. This system does not use ground side switching.
Answer B is wrong because if the control switch shorted to ground, the fuse would blow.
Answer C is correct because if the switch contacts stuck closed, there would be a continuous circuit and constant pump operation.
Answer D is wrong because there is no separate relay in this system for the wiper washer pump motor.

Question #82
Answer A is wrong because while the resistance of the heater may change slightly when it warms up, it will not be enough to causes significant changes in amperage draw readings.
Answer B is correct because it is normal for some heated mirrors to draw a large amount of current during the warm-up phase and then taper down in normal operation.
Answer C is wrong because only technician B is correct.
Answer D is wrong because one of the technicians is correct.

Question #83
Answer A is wrong because the gauge pictured cannot be an ammeter. The scaling shown will not accommodate an ammeter.
Answer B is correct because this picture shows a dual air pressure gauge.
Answer C is wrong because only technician B is right.
Answer D is wrong because one of the technicians is correct.

Question #84
Answer A is wrong because a completely discharged battery will show a reading of around 1.120 or less.
Answer B is wrong because a 3/4 charged battery would show a reading of around 1.225.
Answer C is correct because a reading of 1.20 would indicate a battery that is about half-charged.
Answer D is wrong because a fully charged battery will show a reading of around 1.265 at 80°F.

Question #85
Answer A is a good choice, because checking for proper power at the source is always a good first step in any troubleshooting process. However, since technician B is also correct, answer A is wrong.
Answer B is also a good choice, because faulty wiring can definitely cause intermittent operation. However, since technician A is also correct, answer B is wrong.
Answer C is correct because both technicians are right.
Answer D is wrong because both technicians are right.

Question #86
Answer A is correct because an open circuit at circuit breaker #4 will prevent both the heated mirror and the rear defogger from operating.
Answer B is wrong because a blown #1 fuse will cause only the heated mirror to not function.
Answer C is wrong because an open circuit between the heated mirror and ground will cause the condition described.
Answer D is wrong because an open circuit between the timer relay and the #1 fuse will cause only the heated mirror to not function, which was the original complaint.

Question #87
Answer A is wrong because a circuit breaker is designed to be replaced only if it fails. When a circuit overloads in operation, it will open. It will later reset when it cools.
Answer B is wrong because a fuse should only be replaced if it fails.
Answer C is wrong because neither technician is correct.
Answer D is correct because both technicians are wrong.

Question #88
Answer A is correct because the current for the blower motor does not have to pass through any resistors when the motor is in the high-speed position. The current from the switch bypasses the resistors.
Answer B is wrong. Even though the resistors are wired in series, the current bypasses the resistors when it is in high-speed position.
Answer C is wrong because only one technician is correct.
Answer D is wrong because one technician is correct.

Question #89
Answer A is wrong because the ignition switch does not require a large fusible link.
Answer B is wrong because a PCM also does not have large current requirements. PCMs are usually fused separately.
Answer C is wrong because an instrument panel also does not have large current requirements.
Answer D is correct. The largest fusible link will normally be found near the power source, either at the battery or starter solenoid terminal. This is where the majority of the truck's electrical current requirements originate.

Question #90
Answer A is wrong because while it performs the same job (circuit protection), it cannot be reset like a circuit breaker.
Answer B is wrong because while the wire separates in half during an overload, the insulation does not.
Answer C is wrong because fusible links are not designed to replace fuses. Fuses are much easier to replace, and are considered far more convenient.
Answer D is correct because it is the purpose of a circuit breaker to open a circuit while maintaining insulation integrity so as not to cause any further damage.

Question #91
Answer A is correct because this picture shows a voltmeter measuring the output voltage of the charging system (which is also battery voltage).
Answer B is wrong because to perform a positive charging cable voltage drop test, one voltmeter probe would be at the "battery" terminal on the starter solenoid, and the other probe would be at the output terminal of the alternator.
Answer C is wrong because to measure the voltage drop of the charging ground circuit, one would probe at the alternator case with one lead, and at the battery negative terminal with the other lead.
Answer D is wrong because to measure starter operating voltage, one would probe at the starter, not the alternator.

Question #92
Answer A is wrong because to perform a starter current draw test, the amp clamp would have to be placed around either battery cable.
Answer B is wrong. Even though the battery is being loaded down with the carbon pile, the arrow in the picture specifically shows an amp clamp measuring alternator output. The battery is being loaded to force maximum output from the alternator.
Answer C is correct because the arrow shows an amp clamp being used to measure the current output through the alternator output wire.
Answer D is wrong because parasitic battery drain tests are not done using an amp clamp. A much more sensitive current measuring device (a DMM) is needed for this test.

Question #93
Answer A is wrong because a pulley can be reused if it is in good condition.
Answer B is correct because alternator brushes are considered wear items and should always be replaced during a rebuild.
Answer C is wrong because technician A is wrong.
Answer D is wrong because technician B is right.

Question #94
Answer A is wrong because the battery cables are not considered part of the control circuit; rather they are part of the high amperage starter draw circuit.
Answer B is wrong because the starter motor is not considered part of the control circuit. It is another high amperage draw component.
Answer C is wrong because the starter solenoid high-current switch contacts are not part of the control circuit; rather they switch the high-current circuit from the battery to the starter motor itself.
Answer D is correct because the magnetic switch is part of the starter control circuit. This includes anything between the key switch and the starter solenoid, not including the switching contacts.

Question #95
Answer A is correct because the schematic shows both heater and AC controller components.
Answer B is wrong because an open in circuit #50 (1 BRN) would prevent the AC clutch relay from operating.
Answer C is wrong because only technician A is correct.
Answer D is wrong because one of the technicians is correct.

Question #96
Answer A is wrong because all 12-volt systems used on trucks use DC current for motor power.
Answer B is wrong because installing a relay between the blower motor resistors and the motor would defeat the purpose of the resistors, which is to decrease the speed of the motor.
Answer C is wrong because neither technician is right.
Answer D is correct because both technicians are wrong.

Question #97
Answer A is wrong. The blower motor will operate on the high-speed position because the resistors are bypassed.
Answer B is wrong because removing the resistors will create an open circuit in the low and medium speeds. Since there would be no current flow, the fuse would not blow.
Answer C is correct because with the switch in the high-speed position, the resistors are bypassed and the motor will operate normally, but only in high speed.
Answer D is wrong because with the resistors removed, current cannot flow from the low-speed switch position to the motor.

Question #98
Answer A is wrong because test lamps cannot measure resistance, only ohmmeters can.
Answer B is correct because jumper wires can be used to temporarily supply power to circuit breakers, relays, and lights to check for proper operation.
Answer C is wrong because only technician B is correct.
Answer D is wrong because technician B is correct.

Question #99
Answer A is wrong because any connection that tests more than 100 millivolts voltage drop should be repaired.
Answer B is wrong because any voltage drop test that produces over 100 millivolts at any connection means that part should be repaired.
Answer C is wrong because both technicians are wrong.
Answer D is correct because neither technician is right.

Question #100
Answer A is correct because an open between the ignition switch and the window switch will cause only the window switch to not operate the motor.
Answer B is wrong because an open in the window switch movable contacts will prevent voltage from the master switch to pass through it and on to the motor.
Answer C is wrong because an open in the master switch ground wire will cause no motor operation when the master switch is depressed.
Answer D is wrong because a short to ground at the circuit breaker will result in a blown fuse with no motor operation at all.

Question #101
Answer A is wrong because you should never jump across a fuse or circuit breaker with a jumper wire. A short to ground somewhere in the circuit will cause an excess of current to flow and may damage other wiring components in that circuit or elsewhere.
Answer B is wrong. While this may be an option, it is time consuming and likely not necessary.
Answer C is correct because checking for power at the source is one of the first logical steps taken when troubleshooting a system that does not work at all.
Answer D is wrong because measuring resistance at the motors will tell you very little, other than if it had an open circuit. However, this would be done at some point after checking for power at the source.

Question #102
Answer A is wrong because a bad diode (assuming it was open and not shorted) would only cause the window switch to be inoperative, not the master switch.
Answer B is correct because corrosion at the window switch contacts could prevent current flow coming from either switch.
Answer C is wrong because only technician B is right.
Answer D is wrong because one of the technicians is correct.

Question #103
Answer A is wrong because it makes no sense to check only one motor when the whole system is dead.
Answer B is wrong because even if the alternator were not charging, the motors should still operate unless the battery was totally dead. However, at this stage, the windows would not be the only things that were dead.
Answer C is wrong because neither technician is right.
Answer D is correct because both technicians are wrong.

Question #104
Answer A is correct because a fan circuit may be controlled by either hot or ground side switching, depending on the preference of the manufacturer.
Technician B is wrong because a fan might also come on during AC operation, or when coasting downhill with the cruise control engaged to assist in vehicle retardation.
Answer C is wrong because only technician A is correct.
Answer D is wrong because one of the technicians is wrong.

Question #105
Answer A is wrong because there is no dedicated ground wire in this circuit between the switch and the motor. The motor is reversible.
Answer B is correct because even one faulty set of contacts in the window switch will disable the motor both ways, because both power and ground for the motor pass through this switch.
Answer C is wrong because only one of the technicians is right.
Answer D is wrong because one of the technicians is right.

Question #106
Answer A is wrong because an electronically controlled engine does not have a servo motor. The cruise control is an integral part of the ECM.
Answer B is correct because faulty cab cruise control switches could definitely disable the cruise control.
Answer C is wrong because an electronically controlled engine has no mechanical linkage to the fuel pump.
Answer D is wrong because a faulty vehicle speed sensor would also disable the speedometer.

Question #107
Answer A is correct because an engine ECM can automatically engage the fan if a faulty temperature sender reports a reading out of range. This is done to protect the engine.
Answer B is wrong because the fan only operates in cruise control in a downhill mode to assist other vehicle retarders, or normally during high temperature operation.
Answer C is wrong because only one technician is correct.
Answer D is wrong because one technician is correct.

Glossary

Actuator A device that delivers motion in response to an electrical signal.

AH Ampere-Hours, an older method of determining a battery's capacity.

Alternator A device that converts mechanical energy from the engine to electrical energy used to charge the battery and power various vehicle accessories.

Ammeter A device (usually part of a DMM) that is used to measure current flow in units known as amps or milliamps.

Ampere A unit for measuring electrical current, also known as amp.

Analog Signal A voltage signal that varies within a given range from high to low, including all points in between.

Analog-to-Digital Converter (A/D converter) A device that converts analog voltage signals to a digital format, located in the section of a control module called the input signal conditioner.

Analog Volt/Ohmmeter (AVOM) A test meter used for checking voltage and resistance. These are older style meters that use a needle to indicate the values being read. Should not be used with electronic circuits.

Armature The rotating component of a (1) starter or other motor, (2) generator, (3) compressor clutch.

ATA Connector American Trucking Association data link connector. The standard connector used by most manufacturers for accessing data information from various electronic systems in trucks.

Blade Fuse A type of fuse having two flat male lugs sticking out for insertion into mating female connectors.

Blower Fan A fan that pushes air through a ventilation, heater, or air conditioning system.

Cartridge Fuse A type of fuse having a strip of low melting point metal enclosed in a glass tube.

CCA Cold cranking amps, a common method used to specify battery capacity.

CCM Chassis control module, a computer used to control various aspects of driveline operation, usually does not include any engine controls.

Circuit A complete path for electrical current to flow.

Circuit Breaker A circuit protection device used to open a circuit when current in excess of its rated capacity flows through a circuit. Designed to reset, either manually or automatically.

Data Link A dedicated wiring circuit in the system of a vehicle used to transfer information from one or more electronic systems to a diagnostic tool, or from one module to another.

Diode An electrical one-way check valve. It allows current flow in one direction but not the other.

DMM Digital multimeter, a tool used for measuring circuit values such as voltage, current flow, and resistance. The meter has a digital readout, and is recommended for measuring sensitive electronic circuits.

ECM/EDU Electronic control module/unit, an acronym for the modules that control the electronic systems on a truck.

Electricity The flow of electrons through various circuits, usually controlled by manual switches and senders.

Electronically Erasable Programmable Memory (EEPROM) Computer memory that enables write-to-self, logging of failure codes and strategies, and customer data programming.

Electronics The branch of electricity where electrical circuits are monitored and controlled by a computer, the purpose of which is to allow for better and more efficient operation of those systems.

Electrons Negatively charged particles orbiting every atomic nucleus.

EMI Electro-magnetic interference, a condition caused by strong magnetic fields influencing various electrical/electronically controlled circuits, usually causing random and unwanted acts.

Fault Code A code stored in computer memory to be retrieved by a technician using a diagnostic tool for troubleshooting the problem.

Fuse A circuit protection device meant to open a circuit when amperage that exceeds its rating flows through a circuit.

Fusible Link A short piece of wire with a special insulation designed to melt and open during an overload. Installed near the power source in a vehicle to protect one or more circuits, and are usually 2 to 4 wire gauge sizes smaller than the circuit they are designed to protect.

Grounded Circuit A condition that causes current to return to the battery before reaching of its intended destination. Because the resistance is usually much lower than normal, excess current flows and damage to wiring or other components usually results. Also known as short circuit.

Halogen Light A lamp having a small quartz/glass bulb that contains a filament surrounded by halogen gas. It is contained within a larger metal reflector and lens element.

Harness and Harness Connectors The routing of wires along with termination points to allow for vehicle electrical operation.

High-Resistance Circuits Circuits that have resistance in excess of what was intended. Causes a decrease in current flow along with dimmer lights and slower motors.

In-line Fuse A fuse usually mounted in a special holder inserted somewhere into a circuit, usually near a power source.

Insulator A material, such as rubber or glass, that offers high resistance to the flow of electricity.

Integrated Circuit A component containing diodes, transistors, resistors, capacitors, and other electronic components mounted on a single piece of material and capable of performing numerous functions.

IVR Instrument voltage regulator, a device meant to regulate the voltage going to various dash gauges to a certain level to prevent inaccurate readings. Usually used with bi-metal type gauges.

Jumper Wire A piece of test wire, usually with alligator clips on each end, meant to bypass sections of a circuit for testing and troubleshooting purposes.

Jump Start A term used to describe the procedure where a booster battery is used to help start a vehicle with a low or dead battery.

Magnetic Switch The term usually used to describe a relay that switches power from the battery to a starter solenoid. It is controlled by the start switch.

Maintenance Free Battery A battery that does not require the addition of water during its normal service life.

Milliamp 1/1000th of an amp. 1000 milliamps = 1 amp.

Millivolt 1/1000th of a volt. 1000 millivolts = 1 volt.

Ohm A unit of electrical resistance.

Ohmmeter An instrument used to measure resistance in an electrical circuit, usually part of a DMM.

Ohm's Law A basic law of electricity stating that in any electrical circuit, voltage, amperage, and resistance work together in a mathematical relationship.

Open Circuit A circuit in which current has ceased to flow because of either an accidental breakage (such as a broken wire) or an intentional breakage (such as opening a switch).

Output Driver An electronic on/off switch that a computer uses to drive higher amperage outputs, such as injector solenoids.

Parallel Circuit An electrical circuit that provides two or more paths for the current to flow. Each path has separate resistances (or loads) and operates independently from the other parallel paths. In a parallel circuit, amperage can flow through more than one load path at a time.

Power A measure of work being done. In electrical systems, this is measured in watts, which is simply amps x volts.

Processor The brain of the processing cycle in a computer or module. Performs data fetch-and-carry, data organization, logic and arithmetic computation.

Programmable Read Only Memory (PROM) An electronic memory component that contains program information specific to chassis application: used to qualify ROM data.

Random Access Memory (RAM) The memory used during computer operation to store temporary information. The computer can write, read, and erase information from RAM in any order, which is why it is called random. RAM is electronically retained and therefore volatile.

Read Only Memory (ROM) A type of memory used in microcomputers to store information permanently.

Reference Voltage The voltage supplied to various sensors by the computer, which acts as a baseline voltage; modified by sensors to act as input signal.

Relay An electrical switch that uses a small current to control a large one, such as a magnetic switch used in starter motor cranking circuits.

Reserve Capacity Rating The measurement of the ability of a battery to sustain a minimum vehicle electrical load in the event of a charging system failure.

Resistance The opposition to current flow in an electrical circuit; measured in units known as ohms.

Rotor (1) A part of the alternator that provides the magnetic fields necessary to generate a current flow. (2) The rotating member of an assembly.

Semiconductor A solid-state device that can function as either a conductor or an insulator depending on how its crystalline structure is arranged.

Sensing Voltage A reference voltage put out by the alternator that allows the regulator to sense and adjust charging system output voltage.

Sensor An electrical unit used to monitor conditions in a specific circuit to report back to either a computer or a light, solenoid, etc.

Series Circuit A circuit that consists of one or more resistances connected to a voltage source so there is only one path for electrons to flow.

Series/Parallel Circuit A circuit designed so that both series and parallel combinations exist within the same circuit.

Short Circuit A condition, most often undesirable, between one circuit relative to ground, or one circuit relative to another, connect. Commonly caused by two wires rubbing together and exposing bare wires. It almost always causes blown fuses and/or undesirable actions.

Signal Generators Electromagnetic devices used to count pulses produced by a reluctor or chopper wheel (such as teeth on a transmission output shaft gear) which are then translated by an ECM or gauge to display speed, rpm, etc.

Slip Rings and Brushes Components of an alternator that conduct current to the rotating rotor. Most alternators have two slip rings mounted directly on the rotor shaft; they are insulated from the shaft and each other. A spring loaded carbon brush is located on each slip ring to carry the current to and from the rotor windings.

Solenoid An electromagnet used to perform mechanical work, made with one or two coil windings wound around an iron tube. A good example is a starter solenoid, which shifts the starter drive pinion into mesh with the flywheel ring gear.

Starter (Neutral) Safety Switch A switch used to insure that a starter is not engaged when the transmission is in gear.

Switch A device used to control current flow in a circuit. It can be either manually operated or controlled by another source, such as a computer.

Transistor An electronic device that acts as a switching mechanism.

Volt A unit of electrical force, or pressure.

Voltage Drop The amount of voltage lost in any particular circuit due to excessive resistance in one or more wires, conductors, etc., either leading up to or exiting from a load (e.g., starter motor). Voltage drops can only be checked with the circuit energized.

Voltmeter A device (usually incorporated into a DMM) used to measure voltage.

Watt A unit of electrical power, calculated by multiplying volts x amps.

Windings (1) The three separate bundles in which wires are grouped in an alternator stator. (2) The coil of wire found in a relay or other similar device. (3) That part of an electrical clutch that provides a magnetic field.